FROM THE GROUND UP

THE WELTIS IN THE NEW WORLD
— A PIONEER STORY —

MICHAELA WELTI

TRANSCRIBED BY FRANK APPLETON

Suite 300 - 990 Fort St
Victoria, BC, V8V 3K2
Canada

www.friesenpress.com

Copyright © 2018 by Michaela Welti
First Edition — 2018

All rights reserved.

Front cover photo by Frank Appleton

Transcribed by Frank Appleton

No part of this publication may be reproduced in any form, or by any means, electronic or mechanical, including photocopying, recording, or any information browsing, storage, or retrieval system, without permission in writing from FriesenPress.

ISBN
978-1-5255-1918-5 (Hardcover)
978-1-5255-1919-2 (Paperback)
978-1-5255-1920-8 (eBook)

1. BIOGRAPHY & AUTOBIOGRAPHY, ENVIRONMENTALISTS & NATURALISTS

Distributed to the trade by The Ingram Book Company

M. Welti

Photo credit: Welti Family photos.

I would like to dedicate this book to my husband, Max Welti, who taught me endurance and reality.

CONTENTS

1. A Personal Biography of Michaela Welti   7

2. Canada   39

3. Pemberton—From the Ground Up   63

4. Wild Animal Stories—Wolves, Deer, Bear, and Grizzlies   109

5. Felting: an Ancient Craft Redisovered   119

6. Edgewood   129

7. Of Herbs and Health   141

8. Spiritual Values   151

1. A PERSONAL BIOGRAPHY OF MICHAELA WELTI

Swiss Roots

There is a part of Switzerland called Canton Graubünden toward the eastern part of the country. It is an alpine region of the upper Rhine River Valley. The river Rhine is a young stream there, and its source is not far away on Gotthard Mountain. I was born in Disentis/Muster, a village in the country, in the home of my grandparents', on the third of May, 1933. Unfortunately, that was the same year Adolf Hitler became chancellor of Germany. Much of what happened to us in the years to come would be as a result of Hitler coming to power. I was the youngest child of two sisters and three brothers.

My grandparents were Oscar and Marie Hausamann. My grandfather had a PhD in chemistry from Hamburg, and my father followed this line of work. Their house was in a region where the fourth language of Switzerland—Romansh—was still spoken. Most people spoke German or French, some spoke Italian; but here, tucked into the mountains, was a region where the old

Romansh language had survived from Roman times. It sounded somewhat like Italian. These people were of smaller build, with dark hair and eyes. *Italians!* you would think.

My grandparents' house in Switzerland.
Photo credit: Welti Family photos.

The house of my grandparents was built in 1909 with a big front garden. Behind on a hillside there was a forest. It was a villa three storeys high with a veranda and an attached octagonal guest house called the "Pavilion." It was the seventh house in a group

FROM THE GROUND UP

of houses. This village was called Faltcharidas, which comes from the Latin meaning "people who use scythes." It was about a fifteen minute walk from Disentis. At 1,120 metres, it was way up in the foothills of the Alps.

A huge convent towered over the village, established by the Benedictine monks back in the twelfth century. It had paintings on the ceilings. The pictures were ancient. They told the story of when the Mongols invaded northern Italy and the mountain passes of the Alps. The Mongols wanted to attack the monastery, but a vision—an apparition—appeared in the sky. An angel warned them not to attack, frightening their horses, and they turned back. The monastery and village were saved.

The monastery owned a lot of land. They had a college for boys. They were self-sufficient with vegetable gardens and dairy cows. It was a lifestyle that had remained unchanged for centuries. There were no industries in the area. Most of the people had small farms with sheep and goats—wool spinning and hand weaving were common crafts among the women, and I got interested in that, and still practise these arts.

Life in Norway

My father was a professional chemical engineer whose work took him to other countries. In 1936-37 we moved to Norway, where my father had work in a factory in Oslo that made margarine and ice cream. He had his own laboratory to test his samples. The delicious ice cream was called Premier Ice Cream. We lived on the factory grounds in the main house.

I went to the Norwegian school and quickly learned the Norwegian language, even better than German. When I started school, we were all tested for tuberculosis. I tested positive. My parents were worried, and I had to have an x-ray, which showed scars on my lungs. But my mother remembered that I had been treated for whooping cough when I was two or three years old and had had nosebleeds when I coughed. There was an epidemic of this in Europe at about that time. I remember the long winters in Norway with long nights and short days. But I learned to ski and skate, and in the summer we would stay at a farm where we could ride horses, pick berries, and go swimming.

In 1937, my father bought the house in Switzerland from my grandfather, who died in 1938. My unmarried aunts lived there and looked after the house. My brother Jacki, and later my sister Sonja, joined us in Norway. We celebrated Christmas at my father's boss's house in Oslo. This was a Jewish family, but they had a big Christmas tree in the entrance and an elaborate Christmas dinner table with servants and many fancy foods, desserts, and liqueurs.

I also started piano lessons when I was seven. My father played the piano and cello. Often, friends of my dad's would come and form a quartet together, which was really nice. I still have a fondness for music and play the piano.

World War II

In 1939, war broke out between Germany and England, which was allied with Czechoslovakia and France. The Germans occupied France and then Norway. King Haakon of Norway fled and went into exile in England. By 1942, we were experiencing bombing attacks over Oslo, which I dreaded. The siren that gave the alarm was right across the street, and so loud it would terrify you when it went off, especially at night. We had to close off the windows with dark curtains so no light would show outside. As we lived on the outskirts of Oslo, our district escaped the bombs, but we could see explosions in the sky: flak from the anti-aircraft guns. When the Germans occupied Norway, they opened German schools. My

FROM THE GROUND UP

parents wanted me to attend so I would learn to speak German. This was hard for me, especially the spelling. My brother Jacki was nine years older than me. He attended college in Oslo and, among other subjects, learned English. My sister Sonja found employment with the Swiss embassy in Oslo as a secretary and stayed with the embassy after our family left Norway.

There was a shortage of meat at that time in Oslo, so we ate a lot of fish. On the weekends, we would often go into Oslo city for an outing with my parents. We would visit the park where there were many famous sculptures, or go to the port where you could look at the U-boats. My brother Jacki was a teenager then, and very impressed by the U-boats. These submarines were sinking many British and Allied ships as Hitler tried to cut off trade to England. We would see the great success they were having when we went to the cinema and saw the *Wochenschau*—the weekly newsreel from around the world.

In Oslo, they had big craft supply stores for crafts like weaving, felting, woodcraft, and looms. My mom was very interested in knitting, sewing, weaving and embroidery, and taught me these things, too. My dad liked woodcraft, and had many books on astronomy, astrology, and so on. In the city, we would go to the fine cafés where they had assortments of pastries. A banana was a big thing then, because the war had disrupted trade from tropical countries to us in the north. Sometimes, we would go out for dinner. Most of what was on the menu was fish.

My father was very impressed with Hitler, the way he had set Germany back on its feet, got industry going again, and people working after the disasters of World War I and the Depression. My father's praise of Hitler influenced my brothers, and my older brother, Prosper, joined the army of the Reich. However, my mother thought otherwise about the direction in which Hitler was leading us. In 1942, Prosper was killed in Russia, in the battle for Stalingrad. My teacher from school delivered the news. My parents were devastated. He was only twenty-four years old. They sent us his wallet and watch, and a medal for bravery.

Life and Death in Berlin

In 1943, my dad got a new job in Berlin, as a director of "Sunlight Ag." They made Sunlight soap and many other things. The war got worse for us there. There were many more bomb attacks on Berlin. My dad's work took him to Dunkirk, a city on the French coast. The Germans had plans to build big factories there. They had meetings, which my father attended, and he caught the attention of General Heinrich Himmler, who was in charge of these works. My dad heard later that Himmler had asked about him, "Who is the man with the white hair?" and it was explained to him that my dad was the director, an industrial chemist, who designed factories that turned vegetable and fish oils into soap, margarine, and many other things. My dad was informed that the Germans had big plans for industries in the occupied countries, especially the Ukraine, after the war, and he would be very useful.

My dad's work kept him in Dunkirk, while my mom and I stayed in Berlin. We had brought an entire household of furniture from Norway, but it was put into storage because we only had a small apartment. In 1943, Hitler was recruiting more men for his army. Jacki was nineteen by then, and he begged our father to sign the papers so he could go into the German navy. (As a Swiss citizen, he didn't have to go.) My father signed, but my mother was distraught. It would be a long time before we heard from Jacki again.

Photo of Dad   *Photo of Mom*

Photo credit: Welti Family photos.

Now it was just Mom and I in Berlin. I was supposed to start school, but the war intensified. The bombing attacks happened every night; sometimes the alarm went off in the day, and we had to go to the underground bomb shelter every time. I hated it.

There was now a food shortage to add to the constant nighttime bombing. We were deprived of food and in fear of being killed. One day, my mom looked at me and said, "We are getting out of here!" She had a plan to go to her relatives in Silesia, in eastern Germany. But you needed permission from the Gestapo to travel. We had been waiting for a long time—hours and hours—in the Gestapo office when my mom blew her top! "Why are we not allowed to go? I have already lost one son to the Reich in Stalingrad! For what? Look what's happening! You are going to lose the war! I want to save my daughter and myself!" I was trembling. Saying anything against the Reich got you arrested. She risked a lot. But the SS (Schutzstaffel) "hushed her" and said they would give us the travel papers. So we went to Silesia on the train.

Away from the Bombs and Fear

My Aunt Malla picked us up at the train station with a horse and buggy. I was thrilled! We were in the country again, away from the bombs and the fear! I was delighted to meet my grandmother for the first time and, later, another aunt who came from Hamburg. The village was called "Herbstdorf." They lived in a small house and had goats, rabbits, geese, and chickens. We stayed with my aunt and grandma for a while. It was a big, beautiful country with large fields of grain with borders of red poppies and blue bachelor button flowers. My Aunt Malla and I would take bicycles and go to the flaxseed fields for weeding. In the forest, you saw deer, pheasant, plus many berries, elderberries, and mushrooms. I liked it very much there. But we had to find a place to live, and I had to go to school.

My dad would send us letters from Dunkirk. Though he was so far away, we kept in touch.

Mom and I moved to "Patchkau," a small city not far away. It was one-hour walking distance from Herbstdorf. Patchkau is an ancient walled city built in a circle, with high stone walls surrounding two huge entrances on either side, with a marketplace in the middle. The streets were paved with cobblestones. Most of the people were strong Catholics there.

So I started school. The classrooms were big—thirty-five to forty children in a huge room. The teachers were very strict, especially in gymnastics. There was also "the strap" involved, if you did not behave. Mom and I lived as refugees, in one room in a flour mill driven by water from a river that flowed through the city. As the war worsened, it became harder to get groceries. Mom would

FROM THE GROUND UP

stand in line at the meat store for hours, and would come away with very little. But we could have as much flour as we wanted for baking from the flour mill. The mill had a garden with a gardener tending it. I went there often and learned a lot about how to grow different vegetables and fruits. This is where my interest in plants and nature started. I got sick for a while with a cough and cold. Mom worried a lot about me. But the hot elderberry syrup helped me to get well again. From then on, I was interested in herbs.

As the war progressed, we heard less and less from my father, and nothing at all from my brother Jacki. Had he gone into the navy? Had he been killed? My sister Sonja stayed in Norway. She was supposed to be safe there because she worked for the Swiss embassy, and Switzerland remained neutral in the war.

One autumn in 1944, my mom, aunt, and I went by train to visit a sister of my grandmother's, Great Aunt Shotka. They had a farm. They were butchering pigs, and we could get meat and sausages to take home. Shotka did not look like my grandmother at all. She was taller and had long braids pinned up over her head. I was allowed to sleep in her bed, and in the morning I would watch her comb her long hair and pin her braids up. She must have been a great beauty when she was young. The landscape there was prairie-like, and in winter, the winds blew fierce. But they grew wonderful grain and vegetables. In the fields, they grew sugar beets, which were made into syrup, sugar, and animal feed. My mom would bake cakes with the syrup, which stored well.

One morning in the winter, early 1945, we heard sounds like thunder, but we found out that they were from big guns in the distance. The front was not too far away. The Russians were closing in and the Germans were retreating. One afternoon when I was sledding with my friends on a hillside, we saw these ghostly figures coming through the snow toward us. They were starving, with hollow eyes and no boots. Their feet were bound in strips of grey

MICHAELA WELTI

blanket. They had been prisoners of the Germans who had escaped as the Russians advanced. We were afraid of them and ran home. The next day, German soldiers with Russian prisoners went on foot through the streets, knocking on doors and begging for bread. Some prisoners were found dead, left outside the city in the snow. Now we wondered what would happen to us if the Russians came. Mom and Aunt Olly were worried. We would have to leave the city, but where would we go?

One evening when Mom and I were alone, we heard footsteps coming up the stairs. We were very frightened. Was it the prisoners? Or the Russians? But my mom heard something in the sound of the steps and got very excited. She flung open the door, and there stood my dad in his big winter coat and beaver hat! We cried tears of joy to see him. Mom asked him how he had made it through, as it was a long way across Germany from Dunkirk to Patchkau, and many roads and railways had been destroyed in the war. He said they had had to leave Dunkirk because it was being bombarded by the British air force—the Allies were sweeping through France. He was on a train back to Germany when it was attacked by fighter planes "strafing" (shooting) through the roof of the train. They jumped out of the train and hid in the ditches and tall grass. Thank God he did not get hit. For a while, he hitchhiked, trading cigars and tobacco he had saved for rides on ammunition trucks, which were very dangerous. We were so thankful to be alive and together again.

The Long Road Back to Switzerland

My mom and dad decided we had to leave soon and head for Switzerland. We had to get on a train—any train that was going in that direction. Trains were now very irregular, and most were used in the war effort. Also, it was the end of February 1945, and

FROM THE GROUND UP

a very snowy winter. We packed one big suitcase and some purses. We dressed as warmly as we could. I remember I had a lambswool coat my mom had made for me, and she had altered some of my brother Prosper's pants to fit me. We took some of the dark syrup cakes my mom had made, to sustain us through the trip. At the train station we had to wait for many hours. They said only freight trains were going through, and travelling in them was not too pleasant. But we were lucky! Some freight trains were fitted with seats, and we got on one of these. The main thing was to get away before more railways were destroyed and prevented us from going.

Some of the railway stations distributed warm soup to us passengers. After we had travelled for a day or so, my mom looked outside and said, "We were here yesterday!" My dad did not believe this at first, but when he asked at a station, he was told that the train could not get through to the south because the railways had all been destroyed. The Third Reich was collapsing.

Fortunately, since we had Swiss passports, we were allowed to travel through Czechoslovakia to Prague and then on to Austria and through to Switzerland. We had to wait a long time in Prague for the next train. My mom could speak some Polish, and managed to buy some bread and sausage. German Marks were still worth something here. My dad was worried that we might miss the next train by leaving the station, but we made it back all right. We kept travelling through Czechoslovakia into Austria, and came to Linz, where we had to change trains. It was nighttime, and we had to stay in the train station. We were tired out and weary. There were no beds to sleep in, no hot meals. I had to sleep on a table, with Mom and Dad on chairs, heads on their arms, leaning forward on the table beside me.

Early the next morning, we caught a train toward Bregenz on the border with Switzerland. Only a few hours after we left Linz, we heard there had been an air raid on that city that had destroyed much of it, including the railway station.

We had to stay in Bregenz for three weeks to be quarantined for tuberculosis. Some people on the train were sick with the disease. Some restaurants were open, but there was almost nothing to eat. Sometimes, we had to stand in line for ages to get some food. There were no cigars or tobacco for my dad. During the time we were there, we experienced some bombing raids. We entered Switzerland at the beginning of March 1945, weary, hungry, and tired. We were refugees. Swiss people on the train looked at us with pity and gave us chocolate. They smoked cigarettes. Oh, what a relief it was to feel the prosperity and freedom in this neutral country not touched by war. And we had a home of our own to go to—my grandparents' old home in Disentis, where I was born.

At Home in the Alps

Home at last! We entered Switzerland from Bregenz at the end of February 1945. We had lost everything—money, possessions—but at least we had a home to go to: a wonderful villa in the Alps. We came by train to Chur, the capital of Graubünden Canton. There is a major junction of railways and roads here, whence you could travel west to the upper Rhine Valley; to Disentis, where our home was; or to Davos and San Moritz in the east. To the west, the railway went to Zurich and Basel. In Chur, we changed trains to a smaller alpine train that took us to Disentis. We stopped at many villages: Truns, Flims, Ilanz, and others. After Disentis, the train would travel over the Oberalp to Andermatt, and from there, one could descend into the Rhone Valley to Geneva and Lausanne.

I was very excited when we started to get near Disentis. The railway line emerged from the hills, and you could see the house up on the hill. At the train station, we had an emotional reunion

FROM THE GROUND UP

with my aunts, who had been looking after our villa throughout the war. They were unmarried women. There was still snow on the ground, and we walked fifteen or twenty minutes, over the Plazi Bridge, and came to our house. Since I was only four when we left, I did not remember too much, just the big larch and fir trees around the house under which I had a swing. The house had a big garden, with a woodlot and a nice fence around the property.

For me, it was wonderful to discover everything again. It was a sunny day, and the aunts had coffee and buns ready for us. There were so many things to tell each other, the war having stopped all communication. For me, there was a sleigh, which I tried out right away. I was so happy to have a home again and be living in a peaceful country! In the coming days, there was a lot to be organized. I investigated all the rooms in this unique house. It had an attached pavilion in an Oriental style for guests. When my grandfather was younger, he had travelled a lot to the Orient on business, and he had acquired many wonderful things: the lamp in the ceiling was Arabic, carved in brass, and there were side tables carved and inlaid with mother of pearl. There was also a life-sized painting of a woman in the traditional dress of Alsace, the region known as the Black Forest. The windows in the pavilion were of decorative stained glass, as were the windows in the upper living room. There were green shutters all the windows for the winter. Every room had different imported panelling, and there were many paintings and Oriental rugs. There was also a silk prayer rug from the Middle East, hand-knotted and hand-dyed, with Arabic calligraphy. It was draped over the table in the library. Too valuable to walk on! The library contained hundreds of books from all over the world. Paintings of my grandfather and grandmother hung in the dining room, where we had a beautiful oak buffet and table.

We used this room only at Christmas or when we had guests. The veranda was enclosed with windows but was open below. There

19

was a fountain in the garden, supplied from a natural stream that ran winter and summer. There was also a very nice barn/woodshed, and a laundry/kitchen with a wood-fired laundry tub. In this they would boil all the wash, once every three months. It was an old-fashioned way of doing things, but it served us for a year after the war, until Mom bought a modern wringer washer. I remember there was some rationing of butter and other groceries back then.

After the War

In May 1945, the war ended, and at last we had news of my brother Jacki and my sisters Sonja and Carmen. We also contacted my eldest brother, Wolfgang, who lived in Chur. Jacki had been in the navy for a year, then he had been told he had to go into the army, to fight on the Russian front. After having lost his brother in the disaster of Stalingrad, and knowing the Russians were not going to be stopped in chasing the Germans out of Eastern Europe, this would have been a death sentence for him. Like many young men of that time, he deserted the army and went into hiding in a house in a remote part of Austria, where he stayed until the war was over. When he returned to Switzerland, he was imprisoned for a year for having served in the German navy.

My elder brother Wolfgang had also been sent to prison for four years for espionage—he had been sending radio messages into Germany, which was against the law in neutral Switzerland. But at least they allowed him to keep working while in prison. He was a very good graphic artist, illustrating advertising, doing woodcuts, and making maps of areas of Switzerland from the air, using a helicopter to take aerial pictures. Some of his best works came from the time when he was in prison. He was later a great influence on me and my artistic talents.

FROM THE GROUND UP

My sister Sonja found employment with the Swiss embassy in Oslo as a secretary and stayed with the embassy after our family left Norway. But there had been no way to travel through Germany because of the destruction of the railways and roads. Then they had been permitted to travel through Russia. It was a long trip, and they were not allowed to leave the train, but they were treated well. They must have entered through Sweden and Finland, then travelled southwest through Russia until they reached the Black Sea. Here, they had gotten on a ship to Italy, and so reached Switzerland about June or July 1945. It was great to see Sonja again, tanned and good-looking.

My eldest sister, Carmen, had been in Berlin after the war when the Russians occupied the city. The Russian soldiers were lusting after the German women, and one day when she was swimming at the beach, they had come after her. But she was a strong swimmer and had gotten away. Oh, how good it was that we were now all alive and healthy and reunited in Switzerland!

Back to School

Now I had to go to school again. Because I had started school in Norway when I was six, the Swiss school would not accept me in the grade I should have been in, so I had to repeat that grade. The school was one classroom with different grades. Later, I got transferred to another school in a small village called Disla, down by the Rhine River. It was closer to home, and in winter I could ride my sleigh right down to school! I had a good teacher and did surprisingly well. The same teacher taught me French in private lessons. I think the reason I did so well there was that there were only three pupils in my class, and I got a lot of individual attention. Once a week, a Catholic priest came to teach us catechism and religion. We were taught math, geography, Swiss history, and writing. I liked writing, and won some contests. I became quite philosophical, writing about the meaning of classical poems.

At home, we got some chickens and had a garden again. I started keeping angora rabbits, those with the beautiful long hair. The wool from them sold very well to weavers and knitters. I taught myself the piano with help from my dad. He was looking for work again, and eventually got a job with Buss Ag in Pratteln, close to Basel. My sisters and Jacki also got jobs. That left Mom and I at home, and this was the happiest time of my young life.

My Aunt Ollie came to visit from Hamburg, and told us the story of how her husband had returned home after being a prisoner in Russia. When the train arrived, she did not know him because these men looked so awful! They had grey blankets draped over their heads, and their faces looked like skulls! But he recognized her. She was shocked! She was determined to nurse him back to health again. She got a job in Arosa in the hotel, and after a while, she left to return to Hamburg with enough money and several Swiss watches on her arm, which she could sell at a profit.

One time in winter, we experienced a series of earthquakes that originated in the Rhone Valley. It was scary—trinkets on top of the cabinets began to shake, and the old-fashioned doorbell began to ring. It felt like I was walking on water. I ran outside in my stocking feet, into the snow. I knew it was safer to be outside. In time, the aftershocks subsided.

One summer, to make a little money, Mom took in four young girls from Sweden as student exchange girls from the "Pro Juventute." It was a good time for me to learn how to behave like a lady, since by then at age thirteen, I had become something of a tomboy. I looked after the chickens, geese, angora rabbits, two sheep, dogs, cats, and the garden. In July, I went with my friends up into the alpine meadows, to pick blueberries. The berries were smaller than in Canada, but very plentiful, on short bushes. We would return with buckets full of berries, which were made into jam for the winter. In the fall, we went again, this time to gather cranberries. In July/August we would go and look for mushrooms. They were mostly of the chanterelle type, but there was a pine-mushroom type called "Steinpilz." In July, the alpenrose would bloom high in the mountains. It was a hydrangea type of bush and very fragrant. We also had enzian (gentian) and the flower for which the Alps are famous: edelweiss.

Disentis also had a large hotel/spa with radioactive springs. The waters were good to drink and bathe in, and were said to cure cancer.

Every day in summer, the farmers would round up their goats and sheep, and a shepherd with dogs would take them up into the Alps, where they could graze freely in the alpine meadows for the day, returning in the evening. All the goats and sheep knew their way home. In summer, the cows would also be herded up into the alpine meadows, and would stay there for the summer. There were barns up there where the "herders" milked the cows and made cheese. In the fall when they returned home, all the farmers got their share of a round of cheese, according to how many animals they each had.

Those alpine meadows grew a wonderful variety of herbs and wildflowers. The farmers did all their haying by hand, with scythes, because the hills were so steep that tractors and machinery could not be used. They made the scythe blades by hand, hammering a thin piece of metal so that the edge became very sharp. To dry it, the women raked and tossed the hay all day long. When the hay was dry, it got raked together onto hand-woven linen sheets and made into bundles, which the men carried on their heads to the barn. It was a wonderful, aromatic, spicy-smelling hay.

The grain was also gathered up and made into sheaves. These would be stacked outside the barn, against the wall, on shelves where they would dry for a while, and later got flailed by hand with hand-carved wooden flails to release the grain. This was harvesting as it had been practised for hundreds—perhaps thousands—of years. They also made their own sausages, bacon, and air-dried meat—a regional specialty like jerky but with a wonderful cured flavour. It was carved finely into thin slices, which curled up. The bacon was soaked in brine and then air-dried. They also dried lots of pears and other fruit, and made a special fruit bread with hazelnuts, which kept for a long time. There was also a community wash house and baking oven. Each family would bake enough bread to last for months. Switzerland liked to keep their farmers on the farm and so they were subsidized by the government.

Our Year in Spain

When I was fourteen years old, my dad got a new job in Spain. He connected with a long-time friend from his younger years, when they would go whale-hunting in the Strait of Gibraltar. The friend's name was Rezola, and he was the mayor of San Sebastian, a northern city in Spain. Rezola had industries in Bilbao, and my dad's job was to show them how to make margarine from olive oil. He went to Bilbao first, and Mom and I followed, later. Because we moved, I missed my last year of school, which was Grade 8.

Mom and I travelled to Paris, where we had a few hours to wait, so we visited the cathedral of Notre Dame. Then we took the train south, through France and the Pyrenees Mountains, to San Sebastian, where Dad and Rezola picked us up. We had to stay for a while in a hotel until we could get a house to rent. Bilbao is a port city on the northern coast of Spain, facing the Bay of Biscay. Our house was close to the sea and beaches. There were nice walkways along the waterfront, with many hotels. Mom and I did not have much to do, but we would go and look in the shops, which had wonderful fashions, textiles, and lacework.

When we had our own kitchen and Mom could cook again, we would go to the open markets to shop for fruit and groceries. Mom bought several yards of fabric and a pattern for a dress for me. We had no sewing machine, so we sewed everything by hand. We had Spanish lessons every week from a young woman who spoke German. Through her, we got to know the history of Spain and the lifestyle of the Spaniards. With the Spanish lessons, I could soon speak enough to go and bargain in the market.

The young, unmarried Spanish girls were beautiful, and they were chaperoned when they walked in the city or on the beaches. I loved to go to the beach and swim with my mom. I took up my drawing again, and made a portrait of our Spanish teacher. I also took piano lessons again. This time, I had a professional concert pianist as my teacher, which was very special for me.

Christmas came along, and we were supposed to go home to Disentis to celebrate with the rest of the family. But Mom did not want to go. She loved the mild climate, and said she was perfectly fine to stay alone in Bilbao. (I guessed she wanted to avoid all the work that came with the festivities, as well as the snowy winters!) So Dad and I decided just the two of us would go. We took the train from Bilbao to Barcelona. It was an old-fashioned train with a steam locomotive. We travelled through the night. I was awakened by the train swaying a bit—it was going very fast. The moon was bright, and when I looked outside, we were travelling through olive orchards. We did not have enough time to visit the beautiful city of Barcelona. We had to take a plane to Geneva, where we stayed overnight in a hotel. We had dinner and ordered chicken and a bottle of wine. But we found that the wine in Switzerland was not as good as the wine in Spain. In Spain, we could go to the winery, which had huge barrels of wine. There they would fill up your empty bottles for only a few cents per bottle. It was fuller and sweeter-tasting wine compared to the Swiss.

In the evening, Dad would tell me stories from his younger years when he was in Paris. He knew a few artists who became famous. He would clean their brushes and help organize their studios. We had paintings by some of them in our house in Disentis, some from A. Nador, whose name was noted in the catalogue of the Louvre. The following day, we took the train through the Rhone Valley, and came to Disentis. There, we had a fine Christmas celebration with the family, but I never forgot this trip with my dad, because it was so special and personal.

We stayed one year in Spain, and my Dad's work was successful in creating margarine out of olive oil. Then we returned to Switzerland.

Apprentice Milliner

Now came the time when I should be learning a profession, and I wanted to use my artistic talent. I wanted to learn dressmaking, and be a fashion designer. But at that time, the dressmaking courses were all full, so I studied to be a milliner. The school for textiles was in St. Gallen, and the course was only for females. The apprenticeship would last for two-and-a-half years. We learned everything about textiles and fibres, and some more about running a business, bookkeeping, and so forth. It was strict learning, and by the second year, we were working on consignment, and earned wages, which got paid out when we were finished the course. This gave us a good start in life. I stayed with my sister Sonja, who was now married and living in St. Gallen. During summer vacation, I went home to Disentis.

One time, a friend from school and I decided to go hitchhiking into the Italian-speaking Canton of Tessin and visit the city of

FROM THE GROUND UP

Lugano. We set out over the Lukmanier Pass on a nice summer morning. Hitchhiking was quite common then and pretty safe. We had only five francs each, and one change of clothes, some bread, eggs, and bacon in our backpacks. We also had a little stove, which was heated with Meta tablets. My friend Irena had a friend who worked in a hotel in Lugano, where we could stay overnight. We got some rides, but often had to walk for some time. In the evening when it got dark, we thought we might stay in a grape orchard as it was very warm weather, but along came some people who were singing and drinking, and we felt uncomfortable sleeping there, so we found a youth hostel where we stayed overnight. The next day, we came to Lugano. The first thing we did was find a spot by the lake to go swimming. Later, we bought some peaches and other fruit for breakfast.

We sat at an outside café drinking Perrier and waited for Irena's friend to finish his shift at the hotel. He took us out for dinner and, later, dancing. It was a lot of fun and very adventurous. We got to stay one night in the hotel in a back room. In the morning, we used our little stove to cook our bacon and eggs, with bread. We laughed a lot. We were young and free with nothing to lose. It was a good time and safe to hitchhike. We made it safely back, but when we were close to Disentis, it got dark, and fewer cars came our way. So when a Volkswagen Beetle came, I stood in the middle of the road, desperate to stop it. It stopped, and we piled into the back of the overloaded car, and went very slowly down the road to Disentis. It was completely dark when we arrived, and when we walked up to the house, it was locked and my mom was not home! But I knew a secret way in through a back window. We were both tired and sunburned and slept till almost noon the next day. We were delighted with our adventure. It was a trip I will never forget.

*

In the following years, I visited Lugano twice more—once with my sister Sonja, then again with my cousin Hubert from Germany. We took bicycles, which had only three gears, so we had to get off and push them on steep stretches. The year I went with Sonja was the last year of my apprenticeship as a milliner. When we came home to St. Gallen, my sister came down with polio, which she must have picked up on the trip. There was an epidemic at the time. My mom had been babysitting Sonja's children while we were away, and got concerned when Sonja began showing flu-like symptoms. She called the doctor right away, and he diagnosed early polio infection. She was treated, and thanks to the early detection, had no lasting damage. I graduated from the textile college that year, and got a job as a milliner, but it was not to last long.

Paradise in Brazil

In 1950, my dad, now seventy years old, got a new job far away in Brazil. How this came about is an interesting story. It goes back to when Dad was a young apprentice, learning his profession in Hamburg in the late 1800s. My grandparents lived there, and my grandfather was the director of a big factory making soaps, margarine, and other products. They took on apprentices, and one of these came from Brazil. His name was Wetzel. He was very young, and the other apprentices used to laugh at him a bit because of the way he looked, his oddly accented German, and his having come out of the jungle to learn how to make soap. Then, so many years later, he and my father met up again in Pratteln/Basel. Apparently, Wetzel's parents were German immigrants to Brazil in the 1800s, settling down in the state of St. Caterina. His mother started to make soap, which she sold in her own store. It was a popular item in this remote place, and she decided to send her young son to my grandfather's factory in Hamburg to learn the business of soap-making. That was how my father made friends with him; they were both apprentices together. Now, more than fifty years later, they had met up again. Wetzel had built a soap-making business in Joinville in the state of St. Caterina, and he wanted my dad to go and work for him. So Dad left in 1950 for Brazil. My mom was going to follow him within a year.

My brother Jacki immigrated to Australia in 1950 after he got married.

I went to Chur and joined my brother Wolfgang in his studio to learn graphic arts. It was a good time, and I learned a lot, but it did not last too long. My parents wanted me to come to them in Brazil, and sent me money for the fare. I was to leave from Hamburg by ship in the summer of 1953. The trip was delayed until August. I was staying with my Aunt Olly and Uncle Gerhard in Hamburg. They showed me around, and I met some of my other maternal aunts, and other relatives from my mother's family. I also got to know my cousins Wolfgang, Hubert, and Eberhard. It was a fun time. Eberhard had visited us in St. Gallen. Later, he got married to a daughter of the Bayer Aspirin family. He got very rich and bought real estate in the Mediterranean islands.

Michaela in Joinville, Brazil
Photo credit: Welti Family photos.

My ship left Europe from Amsterdam in August. It was a freighter, which was allowed to take a limited number of tourists. We passed through the canals of Amsterdam until we came to the open sea. When we came to the Bay of Biscay, we had some rough weather. Dishes in the kitchen slid and clattered to the floor in pieces. The passengers were mostly young people, and almost all of them got sick. Only I and a girl my age did not get sick, and we became friends.

At last, we came to the open sea again, and things calmed down. From then on, it was smooth sailing to Rio de Janeiro. We had a short stop in the Canary Islands for about two hours. My new friend and I went to the beach to swim, and thought we were in paradise—the water was so warm and clear, azure blue in colour, with the sand so white! On the ship we had plain, but good

FROM THE GROUND UP

food—lots of beans and fish. The passengers ate together, seated at long tables. Mostly, my friend and I spent time on deck, lying in the sun, tanning ourselves. The cabins had bunk beds. Males and females slept separately. Sometimes in the evening, there would be a dance.

We could often see dolphins, racing the ship. One day, a school of flying fish landed on the deck. At night, it was interesting to watch the big fluorescent jellyfish and other luminous creatures in the water. Many days followed without sighting any land. Sometimes, we played chess, which I was pretty good at. Mostly, I would beat my male opponents, but they didn't like that! As we crossed the equator, the day and night hours equalized. We saw fewer stars than we were used to in the northern hemisphere, only the big Southern Cross dominated the night sky. It was so warm, we spent almost all day and night on deck. As we got closer to Rio, we saw some islands. They were the islands of Fernando de Noronha. The Brazilians used them as a penal colony. Because the waters around them were infested with sharks, the prisoners would think twice about trying to escape.

When we arrived in Rio de Janeiro, I was looking forward to meeting my uncle, who had been living there for thirty years and working in a bank. We had met once, when he came back to Europe, and I was only a young girl. I was standing on the dock wondering what had happened to him when I noticed a man looking at the name on my baggage tag. It was my uncle! He had not recognized me, since I was now a young woman, and I was also tanned from the trip. He had changed, too. After we had a joyous reunion, he showed me where he lived, in a high-rise apartment in Copacabana. Rio is the most beautiful city in the world, with its Sugarloaf Mountain and wonderful white beaches. They speak Portuguese there, which is related to Spanish.

The next day, I boarded a small plane to fly south to Joinville. We flew over a lot of jungle. There were clearings where there were farms with red earth. The woman beside me said many of the settlements were of Japanese people who had come to Brazil. There were immigrants from all over the world. When we came to Joinville, I found there were quite a few German immigrants. This city is not far from the Atlantic Ocean, and when it got unbearably hot in the summer, we moved to a place by the ocean where it was cooler. I remember the papaya trees in the garden and the freshly squeezed sugarcane on the beaches. I did some more drawings and painted some African-Brazilian people. I read a lot of books on Brazil and made a good friend, another German girl.

My dad liked the warmth of the tropics a lot and was happy there. He had plans to buy a car, and I was to learn to drive. We wanted to explore more of Brazil. But it all came to an end sooner than we thought. Mr. Wetzel died suddenly in Switzerland, where he had gone often for medical treatments. His son took over the business, and wanted my dad to stay, but Dad felt it was time to return home. In the fall of 1954, we started the journey back. While I was there, I had visited Rio on a holiday, gone up Sugarloaf Mountain, and spent a lot of time sightseeing and shopping. There are many African ethnicities there. They have businesses in Rio, and will sit beside you in the streetcars. It was nice to see there is no discrimination against them in Brazil.

On our trip home, we were on a French ship called La Bretagne, which had excellent food. We stopped in El Salvador. Its capital is San Salvador, but back in 1954, this city on the edge of the Atlantic was called Bahia. Bahia was famous for having 150 Catholic churches. The city was located on a steep hill, and there was an elevator you could ride from the harbour to the upper city. We were given a few hours to visit, and took the elevator. It was very interesting to see another way of life from Brazil's. We saw many poor people and many beggars. They would follow us like a swarm, begging. Some were disabled, some looked almost animal-like.

Photo of the SS La Bretagne (1951)
Photo credit: Welti Family photos.

One day I was looking for a camera shop, needing some film for my camera, and I went into a shop that had no door, just curtains. Since I was some time in there, my mother was very worried that I might have been abducted. Despite this, we were very impressed with this place. I had read a few books when in Brazil about the people and lifestyles of the area. There is a dark spirituality to these people, especially the poor. They practise voodoo there, which comes from Africa.

Return to Switzerland

My dad was a little depressed on the voyage home, because he knew it was probably the last time he would have a trip like this, and a job in his own field. It turned out that he was correct, because Switzerland was, like the rest of Europe, still economically

MICHAELA WELTI

depressed from World War II. But my dad was very interested in astrology, and made a little money from casting horoscopes for people.

Before we got to Genoa, where the ship was destined, we stopped in Dakar, Senegal, in West Africa, where we could leave the ship for a few hours. It was interesting to see the black people in their long, flowing white garments. But, in spite of the heat, they wore woollen socks right up to their knees. We took a taxi, and almost burnt our bums on the seats they were so hot. We had to hold on tight, as they drove like crazy there. My mom and dad bought two tanned baby crocodile skins. Some years later, these became very nice purses and shoes.

Instead of leaving the ship in Genoa, we sailed on, through the Strait of Gibraltar to Marseille, France. From there, we went by train to Switzerland. When we reached Zurich, it was autumn and much colder, than we had been accustomed to, and so we had to search for warmer clothes. The skin on our hands shrivelled from the cold, dry air, which was so different from the hot, humid conditions in Brazil.

Once again, we settled down in Disentis. The following winter, I started to learn weaving and became very interested in this art. I learned all kinds of different patterns. My teacher was a highly skilled woman, who had been an apprentice weaver for four years. I became very good at patterns and made many different fabrics. My dad bought me a big loom, with flying shuttles. One could weave a two-metre-wide fabric on such a loom. From the loom, my mom and I made fine woollen pants for my dad, and fabric for skirts and jackets for ourselves. My dad was very pleased with our work, and said to me, "With your talent, I don't think you will ever be hungry in your life." I also started to do hand-painted ceramics and later showed them at an art exhibition in Zurich, where I sold ceramics and hand-woven articles. I stayed with a friend, Erica, who I knew

FROM THE GROUND UP

from apprentice classes at college. I made some money. But not enough to make a living.

At that time, a lot of young people talked about emigrating, or going to England for a year to learn the English language. In Israel, they had started the kibbutz system, and were looking for young people to go there to learn agriculture. After the devastation of World War II, people had an urge to improve their lives and build up their businesses. My sister Carmen and her new husband were in the construction business and had lots of work.

Canada Beckons

I had read a few books about Canada, especially stories about pioneer women who were making a living out in the wilderness. I liked nature, plants, and gardening, and could picture myself on a farm in this new land. My brother Jacki, in Australia, and my sister Carmen had both found their partners through advertisements in magazines, and their marriages had worked out well. Somebody suggested I should try this, too. It was one way to get out to one of the newer countries like Canada or America. I had seen a bit of the world, having lived in Norway, Spain, and Brazil, and knew that was the way to learn the language, culture, and way of life in a country. And so I was brave enough and adventurous enough to try a jump into the unknown. Either I was to stay in Switzerland and find a job, or emigrate.

Finding My Life Partner

In 1956, I took a chance and took out an ad in a magazine called *Sie und Er,* or "*Him and Her,*" in English. At first, I said I would like to emigrate and to correspond with people in new countries. But the editor wrote that they would not accept just an invitation to correspond. To be in their magazine, it had to be with the intention to marry. So I had to change this, thinking I had not much chance of finding a partner, especially if I stayed in Disentis.

The response to this ad was unbelievable! I must have gotten about fifty letters from around the world. I had promised to answer all letters, and it kept me busy, writing to men in New Zealand, America, Italy, some African countries, Canada, even Switzerland. Some sent pictures. I met two Swiss men in person, but then came this letter from Max Welti in Canada. What he wrote was so plain and honest. He said he had lost most of his savings in a bad business partnership, but not the courage to begin again. He was Swiss-born in Aargau Canton. He had immigrated to Canada in 1950, and was now working in Port Alice, BC. He was staying in a work camp. A Swiss friend there had given him the magazine where he saw my ad. From then on, we related our lives to each other in many letters.

Max was an orphan who had been born in 1926 in Leuggern, Kanton Aargau. He had two sisters and three brothers. His parents had both died in their thirties of tuberculosis, leaving their very young children. The children had been distributed to other families in the community, some of them quite poor. But the Welti children kept in touch with each other over the years, and the boys each learned a trade. Max went to agricultural college for two years, and it was his knowledge of farming that had brought him to Canada. He had worked in Ontario and Manitoba before coming to British Columbia, where he had worked in the mines in Wells, Kitimat, and Port Alice. He promised me he would come to Switzerland, meet my parents, and if things looked good, we could get married.

But things turned out a little differently. He was laid off temporarily from his job, and went to Vancouver to look for work. We continued to write to each other every week. We exchanged

FROM THE GROUND UP

photographs and got to know each other. At last, we decided I should come to Canada as a bride-to-be, which would help my immigration application. So we got engaged. Max wrote to my dad that he was sincere in wanting to marry me, would take care of me, and if things did not work out, he would pay for my ticket for a safe return home. He also gave me the address of two of his brothers in Switzerland, and encouraged me to visit them, which I did.

I had a friend, Ursula Truaich, whom I had known from elementary school. She had become a children's nurse for a family from the Peugeot car company in France, and had her own Peugeot. So we drove over to Langenthal/Bern and visited Hans Welti and his wife, Ruth, who had just had their first baby. We celebrated the baptism with them, and I became the godmother. I was being welcomed into the Welti family without having met my future husband! But my friend Ursi liked Hans very much, and said, if Max was as charming and good-looking, she would marry him right away! Afterward, we drove over the border into France, to one of the vacation houses owned by the Peugeot family, where we stayed overnight. The next day, Ursi took me to Basel, where I took the train back to Disentis. The other brother, Karl Welti, I visited also, and met his wife, Alice, and their two little boys, Uli and Rolph. This was in the autumn of 1957.

Max had bought two rings, one for himself and one for me. He sent me the ring and the money to fly to Vancouver. Also, I had to go to the doctor for an examination, which was part of my being accepted into Canada. There was also another examination in Bern. And then I was accepted. I was to leave from Zurich on November 25.

I went with my parents to Zurich and we stayed overnight. Our parting the next day was a very emotional time. I was close to tears as we sat at the dinner table. I knew this was a big moment in my

37

MICHAELA WELTI

life, and that it would be a long time before I saw my parents again. (I never did see my dad again.) I remember my dad stretching out his arms to give me a big hug, and my mom fighting back her tears. I also was struggling with my emotions, but I knew if I let go, I would just melt. When I boarded the plane, I looked back to see my parents waving from the platform of the airport building. It was hard for me to keep it together.

But despite the pain and sadness of parting with my parents, I knew and felt inside that everything would turn out fine. I had a dream where I was standing on the edge of a body of water with many little islands, and somebody, who I knew must be Max, was reaching out his arms to help me over. I understood the meaning of the dream was that now I was to enter a new era in my life.

2. CANADA

And so it was! I arrived in Vancouver on November 25, 1957. We flew to London first, then on to Montreal, where we changed planes. I had a window seat and could see the landscape below. We flew over many lakes and the prairies. There was a short stop in Edmonton, and I was astonished to see people wearing thick parkas with fur-lined hoods. It must have been very cold there! As night approached, I saw the Rocky Mountains bathed in moonlight. I arrived in Vancouver near evening.

Max was there! He was standing by the doorway where all the passengers had to pass, and looked a little pale in the face because of all the excitement. We recognized each other right away, and had our arms around each other at last! I had a big suitcase and a handbag. Max took the suitcase and we boarded a bus from the airport to the west end of Vancouver. Max had rented an apartment on Bute Street. We had to walk a few blocks from the bus stop. We were holding hands, looking at each other and laughing.

I would like to mention that, in the year when we wrote letters to each other, we had exchanged many thoughts and philosophies about life—thoughts about religion, music, art, sex, and marriage. We were honest and realistic about it. It was good that way. We had time to get to know each other. Better perhaps than if we had met and seen each other in Switzerland, where our families and sex might have affected the relationship.

The apartment was small, with a small kitchen, living room, bathroom, and a bed that folded up into the wall. There was no table or any chairs as yet. Max had bought dishes and bed sheets, cutlery and a few pots. We ate by using Max's trunk as a table. In a few days, my trunk and other boxes arrived by ship. We were poor, but it did not matter. We were happy and in love. We paid $65 a month for the apartment. From there, we could easily walk to downtown. We were close to Robson Street, which became a favourite place to go because of the many European-style shops.

I had immigrated to Canada as a bride-to-be, and one of the conditions was that we had to get married within ten days of my arrival. Max introduced me to some of his friends. There was Bob Tomasic and his wife, Rosemarie; also Hans and Anneliese, who were German; and Eric Mettchete, who was Irish. We got married in a civil ceremony on December 6, 1957 in Vancouver, with Bob and Rosemarie as witnesses. After this, we had a fine steak dinner with champagne for about $20. Then Bob and Rosemarie drove us through Stanley Park and showed us around Vancouver.

Wedding Day December 6, 1957
Photo credit: Welti Family photos.

Married and Starting a Family

Max had a job as a welder. We spent our honeymoon in the small apartment. We went walking a lot, went window-shopping, went to Eaton's and Hudson's Bay department stores, Stanley Park, and often visited our new friends. I had brought an Elna sewing machine from Switzerland and sewed my own wedding dress, made of navy blue silk and fine wool. Not too long after, Max lost his welding job and had to go on unemployment insurance. I went to evening school to learn English. Every night, we read the newspaper in bed together to improve my English. The English language was hard for me at first, and I learned mostly by listening to the radio when Max was at work. We bought a second-hand dining table and four chairs. The table still exists in my studio in 2016.

Max and I were both very creative, and so we bought some thick wool yarn at Eaton's, along with some canvas, and began to hook a big rug, nine by twelve feet, which I designed. We bought the materials on monthly payments. Later, we traded in my Elna sewing machine for a Super Elna machine, also bought on monthly payments.

Max spent many days looking for work, but could find nothing. The help wanted section in the papers was so small. There were no jobs, and the economy was bad. Then I got a job doing laundry in a lodge. By this time, I was three months pregnant. I got another job sewing, but did not like it. Then I hit on the idea of putting an ad in the paper: "European dressmaker would like to sew in my home or yours." The response was good! Sometimes, I got called to other homes for the day. I learned more English this way, and got paid six dollars a day. We could live on this, pay our debts every month, and even go to the movies on weekends, which we both enjoyed. I liked Vancouver a lot; the city reminded me of Rio de Janeiro.

Farming in BC

Max noticed in the papers that jobs were opening up again in the mining camps, like Port Alice and Kitimat, but he thought the camps were no place for a young married woman. There were too many men there, and few women. He then answered an ad in the paper. They were looking for a herdsman to milk cows and do field work in Chilliwack, BC. We would get a house, free milk and electricity, no rent, and a $150 per month.

First, Max went alone for a week. Then I went out, and we moved into the small house in June. There was not much furniture, and there was some painting to be done inside. The farm was located on River Road by the airport, and we could walk into downtown Chilliwack. In winter, we had a wood stove to cook on and to keep the house warm. But in summer, it was too hot for the stove, and we only had a two-element electric hot plate to cook our meals. The summer of 1958 was very hot, and the small house got very warm, especially in the evening and at night.

There were many blackberries growing wild around there, and on Sundays, when Max had the day off, we would go picking berries and made jam. The hot weather continued into the autumn, and on September 28, our daughter Sonja was born. My mom had sent me a big box of baby clothes, all hand-knitted in Switzerland. I was so glad to get them, and it moved me so much that I became homesick. I wished I had my mom by my side to help with that first baby. But this was our new life, and the choice had been ours. We were our own family now. Autumn came and then Christmas.

In July 1958, Max's brother Ernst came from California for a visit. He was the oldest of the brothers, and had immigrated to Canada a year later than Max. Max had helped him get a job, also in agriculture. But Ernst was a professionally trained chef, who had worked in Lake Louise and Vancouver, BC. And so later found work in Palo Alto, California and settled down. And that is where he lives to this day.

We stayed on the farm until spring 1959, when Max got another job in Abbotsford, also on a farm, but with better wages. But we did not stay there long, either. Across the road was a very scenic farm, belonging to two Italian brothers, the Bassanis, who were bachelors. They offered Max a job as herdsman, which came with a nice modern bungalow-style house, and increased wages.

Photo of Young Max.
Photo credit: Welti Family photos.

I still did some sewing for people, and we managed to save some money. We had previously bought a piano from Eaton's, but it suffered terribly from the moisture in the older houses we had been living in. We thought it would be better in a more modern house. I got pregnant again, and this time it was a boy. Michael (Mike) was born September 17, 1960 in Abbotsford. We lived on Whatcom Road in the Sumas Prairie area. It was quite a distance into the town of Abbotsford—too far for walking. We had been thinking of getting a car, and so bought a 1952 Pontiac. But only Max had a licence and could drive it.

When we were at the Bassani farm, we started to make some furniture—a coffee table and a sofa. I designed a spinning wheel and Max did a wonderful job of it. He got a piece of cedar from

FROM THE GROUND UP

a fence post and carved the spool for the wool. The wheel he cut out of plywood, with a decorative design. Then he glued pieces of thin plywood with the same design to the sides, making the groove in the wheel that held the wool thread. I had that wheel for many years, and spun so much wool on it. I liked to work with wood, too. We discovered we were both multi-talented.

During that time, we made many new friends, my English got better. I could now read books in that language. We could have our own garden now, and grew lots of vegetables. I was amazed at how well vegetables grew in the Fraser Valley. I wanted to learn how to drive the car, but when I applied for a driver's licence, I learned I needed glasses. I thought this was a terrible thing – glasses were not fashionable then. But I needed them for sure, because one day when we were driving through the country, I said to Max, "What a herd of cows in that field!" and Max replied, "You better get those glasses, because those cows are hay bales!" So I got my glasses and then finally passed my driver's test. Now I could take the kids and go shopping! The two children were healthy and good-looking. Besides looking after the household, the kids, and the garden, I still found some time to pursue the arts.

Before Michael was born, I went picking berries in the raspberry fields. I could walk to the neighbouring fields, and took little Sonja with me. I made enough money to buy a basket, diapers, and clothes for the new baby. Max worked hard on the farm. Sometimes, he would have to get up at 3:00 or 4:00 a.m., to do the early milking, and was very sleep-deprived. One time in the summer, the Bassani brothers left for Italy for three months, and Max was left to manage the whole farm himself, milking the cows, making hay, and irrigating. Twice a day, the forty-nine cows had to be milked, using a modern milking machine. The milk went into a big, refrigerated stainless-steel tank through stainless pipes. The milk got shipped every day, and was checked often for bacteria.

MICHAELA WELTI

Max was also responsible for the calving. One time, there was a heifer having calving difficulties. The calf was too big to come out. But Max was very inventive and put a rope around the hooves of the calf, which were sticking out, then put the rope around a post in the barn. I had to step on the rope a few times, with Max pulling on it to take up the slack. The calf was born OK.

We stayed two years at the Bassani farm. Often, a good farm worker gets taken for granted by the owners, and they take advantage of him, loading more work without increasing the pay. As a result, Max was overworked and unhappy. He decided to quit this job.

We moved into a house in the Abbotsford area that was not yet finished. We could stay there rent-free if Max did some work on the house. He went on unemployment insurance for a while. The house was on Clayborn Hill. There were many wild blackberries growing on the hill. We picked them for the canneries and made some extra money with this. All this time, we were dreaming of getting a farm or a piece of land we could call our own.

Then Max applied for and got another job as herdsman on a dairy farm in Ladner, nearer to Vancouver. He was now milking 100 cows twice a day, with wages of $275 a month. Again, we had a garden and a house for free. Then one night (October 12, 1962), there was a terrific wind storm. Hurricane-force winds hit the Fraser Valley and Vancouver area. Max got up early to milk the cows, and discovered that the storm had blown down many trees and taken out the hydro lines. Everything was dark except for flashes and sparks from the downed lines. Fortunately, the farm had a backup generator to run the milking machines. I had never experienced such a storm in my life. It was called "Typhoon Freda." Fortunately, it was an event that never repeated itself.

The idea of getting our own farm was still with us. We looked around and found a small dairy farm in the area of Scott Road in

FROM THE GROUND UP

Surrey. The owners were an older couple who wanted to retire, but when it came to selling the farm to us, their children didn't want them to sell.

A Farm of Our Own

We stayed eighteen months on the Ladner Farm, until we had the opportunity to buy a small farm of 12.5 acres in the Cloverdale area. Six acres were cleared land with a house and barn. It was located on 40th Avenue, and beside it was the famous Kennedy Farm that grew wonderful vegetables on very fertile land. The area was called Mud Bay. It was the Fraser River delta, close to the sea, and if there was a high tide when the river was in flood, the land might have an inch or two of water on it. Many Chinese people owned farmland down there. It was very good soil for growing vegetables. We preferred growing vegetables to milking cows. (Vegetables are very patient, and do not require attention twice a day!) We bought the farm with $800 down and yearly payments. The full price was $8,200. Finally, we had our own property! How enthusiastic we were about it!

I took out books from the library on vegetable growing. We studied them and planned what to grow. Then we found out that we had no "quota" from the marketing board, and that most vegetables required a quota to grow and sell. To get a quota, we would have to start small and work our way up. It would take years. So Max went to the canning companies and got a contract to grow cucumbers and beans. We realized that, to make a living at it, we would need more land. Max then befriended another farmer, Mr. Horsman, who had just bought some land near ours. He proposed a deal to Max, whereby he would let us grow our own crops on ten acres of his land if Max would work the rest of the land, cultivating and planting it with the tractors and farm machinery that

47

MICHAELA WELTI

Horsman owned. We could also use the equipment on our own land. This worked out well.

About this time, we made friends with a Swiss girl who had recently immigrated to Canada. Her name was Margrit, and she was just my age. She had worked in a restaurant in Abbotsford, where she had met a Canadian man to whom she was engaged to be married in a few months. Meanwhile, she quit her job and came to live with us. We became very good friends, and she was of great help to us, looking after the kids while I spent time helping Max in the fields. Now there was a lot of work for me outside, driving the tractor and other jobs. After a while, I found out that I was pregnant again.

We had one of those old sawdust-burner woodstoves to cook on. It had a big hopper on the side for the sawdust. We bought a load of sawdust, put it in the barn, and then filled the hopper with five-gallon buckets from the barn. The sawdust burned very slowly and one could regulate the heat with it quite well. We cooked all our meals on the stove and it heated the house in winter.

We planted seven acres of cucumbers, plus five acres of oats and two acres of beans on our leased land. Everything grew very well—including the weeds! But we were so happy and proud to have our own place, and enjoyed what we were doing. I was glad to be able to contribute, to watch our plants grow, see the creation of the fruit of the earth, and make a living from it. But the weeds kept growing! At first, Max tried to eliminate the weeds between the rows of cucumbers with the tractor as much as possible. But the weeds between the plants had to be pulled by hand. We could not afford to hire anyone, so Max and I went from morning till evening on our hands and knees through the rows, for weeks and weeks. When I would straighten up to ease my back and look around, I would see Chinese families working their fields nearby,

48

with their conical straw hats. It reminded me of reading *The Good Earth*, by Pearl S. Buck. It felt like we were working in China!

Margrit watched the kids at home, but often we took them out into the fields with us, and they enjoyed playing there. They were healthy and happy. Sonja started school in Cloverdale. With the weeding finally done, harvest time was near. What with cooking, washing, and weeding, I had no time to make any clothes for the coming baby. One evening, I took out my spinning wheel and began to spin some wool. "What are you doing?" asked Margrit. I said I had to spin some wool to make clothes for the baby. Then, a few days later, she brought me a big box of baby clothes, given to her by her future mother-in-law who was a member of the Women's Institute, or WI. The WI always had such boxes ready for mothers who had nothing prepared.

"You really deserve it," she said. "You remind me so much of the good housewife in the bible!" I was a bit flustered, but in the end, I was happy to have received such a gift. It reminded me of how serious our situation was. Where was the money coming from? And a third child on the way! Still, we were optimistic. Margrit was such a cheerful person; we laughed often. She also helped us outside in the fields, and since the bigger fields were within walking distance, everyone could come home for lunch. Margrit's fiancé also came and helped us on his days off. One day, they came home for lunch laughing a lot because Max had to tie his shoes together with string: the soles were coming off, and click-clacked when he walked on the pavement. (Max enjoyed the joke.)

We had come so far from Switzerland to the far West of Canada. We had our own little farm and were enjoying our family. But we had no time to do anything but work, work, work on the land, and no money to buy new shoes!

Our First Harvest and Our Third Baby

Now came the time to harvest what we had grown. The Kennedy Farm often hired Chinese workers to weed and at harvest time. We needed a truck to transport the cucumbers and beans to the cannery, plus people to pick the produce. We bought a used 1950s General Motors one-ton truck with wooden box sides. We then made an arrangement with the Chinese to come and pick cucumbers. They arrived early one morning, mostly women, and from inside the house, I heard a sound like a flock of strange birds had landed. It was the Chinese women, about 20 of them, chattering away to each other as they dispersed through the fields! It was a picture to see them with their conical hats, moving through the rows in the field. *The Good Earth* in Canada! They worked for sixty cents an hour—and what workers! Max recalls, "There was one older woman who worked mostly alone, didn't talk much, but picked so fast. Every hour or so she would sit down in the field, have a smoke, then off she would go again, like a machine. Incredible!" The cukes had to be picked small, for pickling, and delivered in sacks to Nalley's processing plant in Burnaby. Every few days, another batch was ready to be picked and shipped.

Before the end of the harvest, our third child, Alex, was born, on September 12, 1963. We were picking the beans in our field by the house when my labour pains started. I took a bath, and then Max drove me to the hospital. He was allowed to stay with me until the baby was born. At that time, they let the woman stay five days in the hospital, and I enjoyed the rest. Max brought me a bouquet of gladiolas from our garden. At home, Margrit was taking good care of the children and the household. When I came home, I was to find the house decorated and a fine "Welcome Home" sign that she had made. I appreciated this very much. I took the baby into the fields with me when I took Max his coffee-break snack. The

Chinese women clustered around and made a big fuss over the baby, praising me for having delivered a second son!

Mike, Michaela and Sonja
Photo credit: Welti Family photos.

The beans got delivered to the Fraser Valley cannery. By the time the harvest ended, we had harvested sixty-five tons of cucumbers. We also harvested two to three acres of oats with a hired combine.

That fall, we went to Margrit's wedding. She married Bill Latter of Abbotsford.

MICHAELA WELTI

Max started working for the Kennedy Farm in the fall of 1963. While Max was working, I took the baby in the truck and delivered the oats to the Surrey Co-op. The next year, Max was working steadily at Kennedy Farm. Now the land on which we had the cucumbers and grain was sold, so we had to rely on our own six-and-a-half acres of cultivated land. We made another attempt at growing cash crops, seeding two acres of onions plus some cauliflower and red cabbage. We also bought some one-to-two-day-old calves, to raise them for sale when they were bigger. But it didn't work out. The calves all got sick because they had not been fed on the colostrum milk, the first milk from the mother cow, which contains special rich nutrition and vitamins to protect the newborn. Later, it became illegal to sell such calves on the market until they had been properly weaned.

Tons of Onions

Once again, we had a Chinese crew to weed the onions. We wanted to start a quota with the marketing board. But there is a lot of processing involved with onions: when they are ready for harvest, they have to be pulled, then laid out in rows to be cured and dried. There were machines for that at the Kennedy Farm, which Max was allowed to use. So we harvested about five tons of onions. Max put them in the barn for further handling and drying. We sold one part of them, the nicest grade. Then we had to wait our turn until all the other onions on the market were sold before they would buy ours. But when we thought we would finally make a bit of profit, there came a bill from the marketing board. They charged us so much for sorting and handling that there was very little profit left! And Max had worked so hard. It was very disappointing. Some neighbouring farms took a few sacks and sold them for us, and we were left with the rest of them.

FROM THE GROUND UP

Halloween came along. Max was in the barn, sorting the onions. We had quite a few kids coming to the door for trick or treat. This was fun for our kids. Then, when we thought the knocking and coming and going were over, there came another knock at the door. I opened the door, and there was a sack sitting on the steps! No face, no hands, no feet. Just a sack.

"Trick or treat?" said the sack. And it began to shake.

I said, cautiously, "Where is your bag?"

More shaking from the bag. Was it laughing? Finally, I realized: it was a person! Max had come from the barn and was laughing so hard he could barely stand.

Then Max emerged from the sack, laughing. "I really tricked you!" he said, but when he saw the look on my face, he ran off down the road with me after him, furious because I had been totally surprised. I never forgot this Halloween trick, but it showed that Max had not lost his sense of humour, despite our worrying situation.

Onions for Powell River

That fall, Max's friend Bob Tomasic visited us. He had quit his job as an architectural draftsman for MacMillan Bloedel and had bought property in Powell River where he'd gone to live with his family. He wanted to go into the logging business and asked Max if he would come and work for him, with the promise of better wages. Max's job at the Kennedy Farm was by then very steady, and although we could live off this, we were not likely to be prosperous. And we didn't have enough land to make a living off that. Max promised he would come to Powell River after he had figured out how to sell the remaining onions with no quota. Bob said he believed the onions could be sold directly to shops in Powell River. So Max put the onions into sacks and loaded them onto the truck,

53

two tons of them, and started out for Powell River. It was quite a journey.

To get there, one had to travel to Horseshoe Bay in West Vancouver, take the ferry to Langdale, drive about fifty miles north up the Sechelt Peninsula to Earl's Cove, take another ferry to Saltery Bay, then drive up the coast to Powell River. Bob promised to meet Max when he got off the ferry in Saltery Bay. So off went Max. It was the first time we had been apart since we were married. But Max had a lot of trouble with our old, overloaded truck on the way. After crossing the Second Narrows (Ironworkers) Bridge, there is a long, steep grade, which brings you to the Upper Levels Highway. There, the truck stalled. Fortunately, Max had a friend in North Vancouver from whom he borrowed some tools. He found that the fuel pump had failed. He managed to get it going again, just enough that by avoiding the big hill, he was able to drive it to a garage where they were able to install a new pump.

He made it onto the ferry to Langdale, but when he started up the Sechelt Peninsula, it was dark and raining heavily, with fog rolling in from the ocean. What with the headlights of oncoming cars dazzling him, the driving became almost impossible. Max went too far over to the side and went into the ditch. He had to walk to find a place with a tow truck, and get them to pull him out. He got out and was able to carry on to Earl's Cove, but when he got there, it was late at night, the ferries were finished for the day, and there were no signs of life, no motel or café or even a phone! He had to sleep in the cab of the truck.

That night, Bob phoned me and said Max had not arrived. We were so worried about him. But when I went to bed that night I thought I would try to reach out to him with ESP. It sounds strange, but even before I left Switzerland, I felt like Max and I had such a strong bond, we could somehow communicate. So I lay

FROM THE GROUND UP

in bed and reached out to him, and then I knew he was all right. Max slept in the truck and was on the first ferry in the morning.

He and Bob tried selling the onions to local stores in Powell River, and got a good reception. They were having some lunch in Bob's car, with our truck parked across the street, when some policemen came up to check out the truck. They asked them if they knew who owned it. Bob and Max said, "No, we don't know!" (It was illegal to sell produce directly to stores except by their regular suppliers, whose onions had come via the quota system.) But our onions all got sold very soon, and then Max had a new job, driving a skidder in the woods. It was something he had never done before. He stayed with Bob and his family, and managed to come home for Christmas.

Meanwhile, I was left alone with the kids, but I could manage just fine. I found I could even change the wheel on the car when I got a flat tire. It was a very cold winter. The water in the fields after so much rain turned to ice, and there was some snow. We had no sleighs or skates, but the kids and I took some cardboard boxes and had some fun using them as sleighs. It was snowing when Max came home, just a day before Christmas. Max had money from his job, and we went shopping, glad to be together again, and celebrated a very happy Christmas.

Now the big rains started. There was so much rain that there was a huge slide on the Hope-Princeton Highway (January 9, 1965), Canada's biggest recorded landslide. Several cars and trucks were buried and four people lost their lives, including Mr. Starchuk, who had sold us our little farm. He drove big transport trucks and was stopped in that area by a small slide when the big slide came down. His was one of the two bodies that were recovered.

Max and Maya, Powell River.
Photo credit: Welti Family photos.

A Visit and Move to Powell River

After Christmas and New Year, Max went back to Powell River to his job in logging. He had never worked with those machines—it was a new experience. It looked like our lives would change again. In January 1965, I took the kids, and we went to visit Max. He had rented a small house, which had been a former boathouse, right above the beach on Lang Bay Road. The kids and I had never been outside the Fraser Valley, so it was a totally new experience—a real adventure for us.

The boathouse was not to be a permanent residence for us. There would be a bigger place available in the spring. It was a foggy morning when we left our farm, but at Horseshoe Bay, the fog lifted, and it was clear all the way to Langdale, a good portent.

FROM THE GROUND UP

There was plenty to see, especially for the kids. We stayed in the little boathouse over the weekend. Outside on the beach, the kids found seashells and crabs. There was a concrete pier there, and at night we went to sleep hearing the waves beating against it. We enjoyed the most spectacular sunsets there. We strolled along the beach until we came to an inlet; walking around it led us to another beach called Palm Beach.

I went back to Cloverdale with the kids, since Sonja had started Grade 1. We all were excited by our experience in Powell River, and we decided to move to Lang Bay Road. We moved on February 8, 1965. Max had come back with the old truck, onto which we loaded all our furniture and belongings. The older tractor we had purchased was picked up later. We put the house up for rent, and in a short time, it was occupied.

In Powell River, later in spring, we were able to move into the storehouse. Again, it was an older building that had been a store with living quarters, but it was liveable. It did not matter to us how it looked, it was the location right by the beach that mattered. We could go swimming any time we wanted, and had a small garden, and some chickens. That summer, the kids took swimming lessons, and there was always something to find or do on the beaches. I even had time to get back to my artwork; I would take out my watercolours and sit on the beach and paint.

Bob Tomasic, his wife, Rosemarie, and their children had property located a little closer to Powell River, on Myrtle Point. They had five children, of a similar age to ours, and we visited each other often. We were happy there, where life was so different than it had been on the farm. It was nice to have more time for the kids and my artwork. We built a small chicken coop with a fence around it, and had a few chickens and a rooster, plus a small garden, which we fertilized with seaweed collected on the beaches. We planted cucumbers among the rocks that lay in a pile by the garden. The warmth of the sun on the rocks plus the seaweed created a fine lot of cucumbers.

Our landlord was Mrs. Barret, an elderly widow. She was very friendly. She lived by herself in a small house on the same property, and owned the boathouse and storehouse. On Sundays, we went exploring around Powell River and as far as Lund, where the road ends. From Lang Bay Road to Powell River was about ten miles. We went shopping there. The pulp mill was there, and there were good stores. We liked living on the coast, with its beautiful scenery, and lovely arbutus trees with orange bark.

We picked huckleberries of the red kind, and yellow raspberries. On weekends we picked salal, a bush that only grows on the coast. It has black, sweet berries and evergreen branches with lovely large green leaves. The berries have always been a staple of the Indigenous diet, and there is a market for the leafy twigs, which are used by florists to make floral displays. We took little Alex with us and picked salal to sell for displays, to make a little extra money, which we used to buy a wringer washing machine. Now I did not have to visit the laundromat whenever I had to do my washing. The salal berries also made very good jam. We liked the climate there: the winters were mild—we only saw snow on the beach once, and it was soon gone. Now we could even save some money, because Max was well paid.

Max's brother Ernst and his wife Trudy and daughter Arlene came for a visit one summer from California, which everyone enjoyed. We also made more friends in the area, playing badminton. We bought fresh-caught fish from the fishermen, who came right to our door, and showed us how to descale and prepare our purchases, mostly salmon and cod. Barbecued fish on the beach was so yummy!

A Year of Accidents and a Death in the Family

That first year, on April 12, Max had an accident at work. He did not come home for supper and I got worried. Then Bob came and said that Max was in the hospital at Powell River!

The skidder had overturned, and in trying to get free of it, he had broken his ankle in two places. He was on crutches for weeks that summer. He was miserable and impatient, but we had the idea of creating a mosaic table for the garden to keep him occupied. When we went to the beach, we collected nice-looking pebbles in a variety of colours. And also at the beach, we found nice pieces of driftwood to make a bench for the table. We cut out a piece of plywood for the base, and nailed the frame around it, cementing the pebbles on it according to a design we had drawn. This kept Max busy so he was not dwelling on his injury.

That summer, in June, my dad died in Switzerland. He was eighty-five years old. I was so sad that I had not seen him again, and had never been able to introduce my family to him. Though we had kept in touch with letters and photographs, we could not afford the trip back to Switzerland.

Eventually, Max went back to work. This time, he was working in Marleny Cove, which is beyond Lund. With the kids, we took him up to Lund to take the floatplane. We saw him jump on the pontoons, board the plane, and fly away. The kids were very impressed! But later that year, Max had another accident. It was late in the summer and I was at the beach with the children when, suddenly and unexpectedly, there was Max! By the way he was walking, I knew something was wrong.

He told me he'd been driving the skidder in the bush when a broken sapling had slammed into the cab, hitting him in the side, in the region of his kidneys. He did not have much pain, but when he saw blood in his urine, he got worried and came home. That night, he was in terrible pain. I phoned the doctor and rushed him to emergency at Powell River Hospital, driving as fast as I could. They found that he had an injured kidney that had caused a blood clot. But it could be dissolved by medication; otherwise, he would have to have an operation. The medication worked, and he got well again. Thank God!

In time, Bob's logging business did not go so well, and he got into financial difficulties. Max stopped working for him, and got another job with Fix Logging, thinning and spacing trees, also he started to learn blasting for road building. He studied blasting techniques, and got his ticket in Vancouver.

One day, I saw an ad in a magazine for angora rabbits in the US. They were a new German breed that produced more wool. I remembered the beautiful wool that came from the angora rabbits we had at home in Switzerland. I bought one pregnant doe; she came by airplane and boat. We picked her up at Powell River Marina, happily munching some hay! From her, I had many more rabbits. What I did with the wool I will mention later.

Following Our Dream to Pemberton

Meanwhile, our dream of owning our own self-sustaining farm had not died. We knew that we needed more acreage than our small farm in Cloverdale. We liked the climate on the Sunshine Coast, and looked around the Powell River area for some land. But it is a very rocky, forested area; the most arable land we found was ten acres. One weekend in the fall of 1966, I saw an ad in the *Vancouver Sun* offering 160 acres in the Pemberton Valley for $16,000, with a down payment of $2,000. That hit the spot!

But where was Pemberton? We found out that there was a new road from the ski resort town of Whistler, but it was not yet

FROM THE GROUND UP

paved. Before this, you could only reach Pemberton with a four-wheel-drive vehicle. Immediately, we wanted to go and see the place. I phoned the real estate agent in Pemberton and made an appointment for the weekend. It turned out that Max had to work overtime that weekend, and was paid double time, so I had to go alone. I had wanted to take the kids, but it would be a long trip for them. Our landlady, Mrs. Barret, volunteered to babysit.

I left early in the morning and took the two ferries to reach Horseshoe Bay, then the road to Squamish and Whistler. The first of the ski-lifts had been built at Whistler. Beyond Whistler the road was dusty gravel. Not too rough. I had made an appointment with Bob Taylor, the real estate agent, who was waiting for me in Pemberton. Closer to Pemberton, the country and road got wilder. The road led through a narrow, winding ravine with a fast-flowing river: the Green River. A railroad ran alongside it, and the road crossed and re-crossed the tracks. At last, I came to Pemberton, crossing the railroad tracks again. On the left was the hotel and on the right the pharmacy, and then the Bank of Nova Scotia. What struck me was that they still had wooden sidewalks, just like in the old Wild West movies!

I found Bob Taylor at the gas station and garage, which he owned along with the real estate office. He was a friendly older man, and with his country clothing and wool shirt with logger's suspenders, he, too, looked like he belonged in the Old West. He insisted he drive our 1952 Pontiac up the valley so I could look around. As we drove, he showed me all the farms and various places of interest. He knew all the names of the families. We drove slowly, and it took quite a while until we came to the land that was for sale. The valley made a slight bend, and on the left side, was a steep mountain; then there were partly cleared meadows on both sides of the road.

Bob explained to me the layout of the land: some of the acreage was on the mountain; a sluggish river called "the Slough" encircled the property on the right side of the road. Most of the farming was cattle, but there were some seed potatoes growing in the valley. I liked what I saw, especially the soil; it looked so fertile. The ferns and dandelions were gigantic! I had never seen such growth. When we left, I looked back at the meadow. I had a feeling of predestination—there, on that spot, would be where the house would later stand! I promised Bob that we would return with Max and the kids soon, in the fall. I returned to Horseshoe Bay and made the last two ferries home.

Now we had to do a lot of planning. We had some savings, but it left us a bit short for a proper start. So we asked Ernst, (Max's brother in California) for the $2,000 down payment. Ernst agreed to give us the money, on the condition that he was to be part-owner in the beginning, until we could get on our feet and repay him. So we went back in October with the kids. They were just recovering from the measles, but felt good enough for the trip. It was getting cold in Pemberton Meadows. The cattle's breath blew clouds of steam. We closed the deal with Bob Taylor, and were now owners of 160 acres of land in the Pemberton Valley. We had to pay $1,000 a year in payments, with six percent interest.

3. PEMBERTON—FROM THE GROUND UP

Building Our First Cabin

It was now the winter of 1966-67. We had a lot of planning to do, aiming to move to Pemberton in the spring of '67. There was no electricity or running water on the property; we had to do everything with hand tools. How could we live there with three kids? We needed a cabin to live in until we could get the house up.

Max had no experience building a house. We had the *Handyman's Encyclopedia,* which we studied intensely. Also, we looked at the ads in the newspapers for pre-built cedar homes and A-frames, but we didn't like the prices. We planned our cabin in a very practical, pre-fabricated way, so that it would be easy and fast to put up. It had a floor plan of 8x16 feet, with one door and one window. The floor, wall, and roof took just twenty-four sheets of plywood, with 2x6 lumber for the floor and 2x4s for the walls. We could cut out the openings for the window and the door with a handsaw.

MICHAELA WELTI

When we had been there in October, we got to know our neighbours, the van Loons. They were an immigrant Dutch family with four kids of a similar age to ours, and lived across the slough from us in a small log house. Max had questioned them about the property we were about to buy. Tony and Nell were very friendly and gave us plenty of information about the Pemberton Valley. Most farmers grew seed potatoes and some also raised cattle.

The Pemberton Valley is interesting geographically. The Lillooet River comes out of the Lillooet Glacier and flows through Pemberton and Mount Currie, then into Lillooet Lake and on into Harrison Lake. Before the valley was settled, the river ran wild through it, and flooding was common when there was heavy rain or snowmelt. To drain the land for agriculture, the river was diked to keep its course to one side of the valley. This left a very fertile, silty loam with few rocks—perfect for potatoes.

The valley ran northwest, ending at Mount Meager and Meager Hot Springs, with Lillooet Falls. One could travel higher up on a logging road, above the falls, where there was a plateau with flat meadows and lots of big huckleberry bushes. We had never seen such amounts of huge huckleberries! The roads up there were white with volcanic pumice stone, which had a foamy texture and was so light that it would float on water. There had been volcanic activity there as few as 2,000 years ago. There was a type of moss growing there with metre-long branches, club moss, and it had some medicinal properties. It likes to grow where there is sulphur in the ground. Groves of huge cedar trees were all around. I had never seen trees of such a size. It took quite a few people, holding hands, to encircle a tree. These trees were 800 to 1,000 years old. One tree had an opening in the bottom of the trunk that a small car could drive through.

The Meager Creek Hot Springs became a favourite place for us, and, later, a popular tourist destination. But sometimes slides

FROM THE GROUND UP

would occur due to unpredictable seismic activity underground. Plans were drawn up to develop the hot springs, but when we moved to Pemberton, the area was totally wild and the road was very rough. Increased logging there gradually improved the roads. It was about thirty kilometres from our farm to the hot springs. Shortly before our farm, one had to turn right and drive over the Forestry Bridge, along the other side of the river, and on to the hot springs. Later, there were many people that drove that bumpy road to the springs. They would party there, and sometimes drink too much, causing accidents and fatalities when they left the road and went into the river.

Meanwhile, Max was still working in the Powell River area, and we stayed at Lang Bay Road, planning our move to Pemberton in the spring. There was an incident worth mentioning around this period that would later influence our lives. Max had acquired a friend at work who was a Jehovah's Witness. This is a global Christian religious sect, and I had heard of them in Europe. This man came over for a visit when Max was recovering from his skidder accident, and they would read the bible together. No, we did not become Jehovah's Witnesses! Max and I had discussed the topic of religion in letters, even before I came to Canada. Max was Protestant and I had been baptized a Catholic. We had both had religious instruction in school—mandatory, in our time. Max had asked in a letter whether I believed in God, and I had replied, "I can see God everywhere in nature." I recalled how I had borrowed a book of bible stories to bring home to read, and was deeply impressed, especially about the life of Jesus and the miracles he did. As for Max, he did not join any church, but never worked on Sundays, just quietly read the bible. For us, the holy bible is the ultimate truth that has prevailed for over 2,000 years. The only way to confront the Witnesses and other Christian sects was to read

65

the bible and interpret it for ourselves. So the bible became the "hub" of our lives.

Toward the end of April 1967, we telephoned Tony van Loon to enquire about snow conditions in the valley, and were surprised when he said there was still plenty of snow on the ground. But the materials for the pre-fab cabin were ready, and we were eager to go and set it up. Max did not want to miss any work, so the two of us left on a Friday afternoon, with Mrs. Barret looking after the kids.

By the time we reached Horseshoe Bay, it was late in the day. We filled our old one-ton truck with gas and headed up the road toward Squamish. We had loaded up the plywood and studs for the cabin, a culvert so we could drive across the ditch, plus flashlights, and a camp stove to cook on. All of it was covered with a tarp. But as we came near Britannia Beach mine, the truck began to swerve and was hard to keep on the road. Max took a flashlight to check, and found one of the tires had a big "balloon" on one side. We had to stop on the side of the road by a little creek. There was nothing to do but stay there overnight. We made room in the back among the building materials, draped the tarp overhead and went to sleep.

It rained overnight and, in the morning when we awoke, the tarp hung down over us, full of water. We pushed up with our feet to get rid of the water, laughing about the impossible ideas we had. It was a Saturday morning and we were not far from Britannia, so we hitchhiked there and found a garage. The mechanic there drove out to take a look at the tire and found it useless. He phoned around to find a replacement, but there was none of the right size available. So Max phoned Eric, an old friend from Kitimat who now lived in Richmond. Eric not only found a new tire, he brought it out to us, filling the whole backseat of his little Volkswagen with this huge tire. We got the tire on, and Eric wanted to come with us to Pemberton to see what kind of property we had bought. He helped us unload the truck when we got there, and I did my best

FROM THE GROUND UP

to cook us a dinner on the camp stove. There were still patches of snow left in the meadows. Eric said he was impressed with the rawness of our wilderness there, and our courage to take on such a project. Then he left to go back home, and told us later how scared he was, driving on the narrow "suicide hill" road to Whistler.

We worked until dark and then wondered what to do with the rest of the evening. In the end, we went over to the neighbours. Tony was home with the kids; his wife was in Squamish Hospital having a baby, their fifth child. We slept in the back of the truck, covered with our hand-hooked wool rug that we had made so long before when we were first married. In the morning, Max went looking for two straight logs of the same size to hold up the floor of the cabin. He felled the trees with an axe, and John, the eldest son of Tony and Nell, came over with their tractor to skid the logs out and helped in cutting them to size. We set up the cabin under some huge cottonwood trees at the back of the property. Time was short, because of the delay we had had with the truck, so we could only get the floor down. We left the unfinished project and returned to Lang Bay Road.

Moving Day!

Our final move to Pemberton came at the beginning of May 1967. We had lots to do before leaving. We butchered our chickens and I canned them in jars—they would serve us for a quick meal in the future. One hen was sitting on eggs and we decided to take her with us, along with the angora rabbits. She hatched the eggs before we moved, so we had a new flock of little chicks. We loaded everything on the truck, which Max drove while I drove the Pontiac with the kids. Our old Oliver tractor was also delivered on a flat-deck truck on the day we moved.

67

For the children, the whole thing was an adventure that would not be repeated. We didn't even have a roof over our heads! After we had put up two sides of the cabin, we put our bed in one corner. Sonja and Mike slept in the truck. Alex slept in the car with some sheepskin to cover him. Later, we rigged up a board over our bed for him to sleep on. The mosquitoes were fierce! We drew our blankets over our heads, but they still found ways to get in and bite us. The kids in the car slept well with the windows closed. In the morning, the mother hen would come and greet us with her cluck-cluck and wander happily around the cabin floor, wanting some breakfast.

We swore that morning that the next night we would have the cabin finished. Sonja and Mike had started school when we were in Lang Bay, and now were eager to go to their new school on the bus with the neighbour kids. It was still cold in the mornings. We worked all day to get the cabin finished, which we did, except for the door. We put up a bunk bed for Sonja and Mike and some shelves for kitchen stuff, and installed our table and chairs. Max had made a cement block with a hole in it for a stovepipe, which he set into the wall to make a stove to cook on and bake in. We had brought a freezer, fridge, and electric stove, but as we had no electricity, we stored these in the neighbour's barn. Max also built an outhouse, and slashed through some bush to make a trail to get water from the slough.

We needed water for drinking, cooking, and washing. Max struggled along his trail with two five-gallon buckets filled with water. He carried them the old-fashioned way, hanging from a pole across his shoulders. But by the time he got to the cabin, half the water had spilled out. So then he slashed a wider trail through the bush so he could drive the truck to the slough, and added a ramp of boards so you could walk out over the water.

FROM THE GROUND UP

We also needed a wood stove to cook on and bake in. So one day he left early in the morning to drive to Cloverdale. We still owned the small farm there, which we rented out. But our renters had not used the old sawdust stove, so Max picked it up and brought it back for us. He took off the sawdust hopper and closed up the side. Now we could use wood to cook and bake. In the evening, we had two old-fashioned oil lamps for lighting.

We had a large zinc-plated washtub, big enough to sit in, for baths. In this, we all had a good "wash-down" on the weekends. This is the system we used: First, kneel down and wash hair. Then, sit down and wash body. Now, stand up and wash legs and feet. The water was warmed on the stove, and thrown in the bush after use.

Building the Big House

In time, we got to know more neighbours and made more friends. We started to plan the house we would build. If a house was within 1,000 feet of the nearest power pole, BC Hydro would install the power to the house for free, provided the homeowner supply the poles necessary to reach the house from the road. Our neighbours the van Loons were the last farm on the road to have electricity, so we had to situate the house close to the border of our property with theirs. We also put the culvert in, so that we could drive into the site.

After we had decided on the place where we would build the house, we sat down to draw up a floorplan for it. I was good at drawing, and made the plan big enough to show every detail. It would be a bungalow-style house on one level. Again, we considered the sizes of the materials from the lumber yards, so that we would not have to do too much sawing by hand.

69

Pemberton House.
Photo credit: Welti Family photos.

The house measured forty-eight by twenty-six feet, with four bedrooms, one kitchen, one bathroom, one living room, and an entrance room (mud room) in the back. It required ten aluminium-frame windows, three doors, 2x4 lumber for the walls, 2x10s for the floor joists, 2x6s for the roof rafters, metal roofing, insulation, boxes of nails, and enough plywood for flooring and sheathing. At that time, we did not need a building permit to start work. We hired a Caterpillar tractor to come and do some ground levelling where the house was to stand. Max made about seventeen concrete blocks for the foundation, mixing the concrete on a plywood sheet by hand. We ordered all the lumber from Cloverdale Lumber on Main Street in Vancouver. It was the cheapest place to buy the wood, and they provided a delivery service. The plumbing and electrical supplies we bought at Beaver Plumbing.

We needed money now, and inquired at the Bank of Montreal for a loan. We had dealt with them in the past, and because we had an excellent credit rating, we could get the maximum personal loan of $5,000, with $160-per-month payments. This loan covered

FROM THE GROUND UP

everything we needed for the house, even boxes of tiles for the kitchen, bathroom, and mud-room floors. I like to mention these prices as compared to their counterparts in 2016, forty-nine years later, when the prices for those materials was five to ten times more, and the quality of the lumber far inferior.

One day, as we were helping the neighbours plant their seed potatoes, the truck with all the materials arrived. It was a big flat-deck, and the package of lumber was so big, it would not slide off the truck. Tony and Nell van Loon could not believe what was happening! But they pitched in to help us unload it and, at last, we could start building! We had nice weather, and soon it got very warm. The mosquitoes were still very bad, especially in the morning and evening. Another neighbour offered to spray the bushes around the cabin with DDT. It helped some, but little Mike got a swollen arm for a while from the spraying. Our boxes of belongings were still lying around, so Max built a lean-to on the side of the cabin.

Drilling for Water

The hauling of water was a daily stressful chore. We inquired about the water situation in the valley, and it seemed almost everybody had a sand-point. This is a pipe with a pointed end, and the point has holes in it to admit water after it's driven into the ground. We had additional lengths of pipe to add onto it as it was driven in. Elmer Hollowang offered to come over with a fence pounder on his tractor to drive it in. We decided to drill close to the house site, where the grass was nice and green. The water has to be tested by boiling for rust and iron, which settles to the bottom of the pan, every twenty feet that the sand-point goes down. They stopped at fifty-two feet, where we found very good clean, cold water. They put a hand-pump on the top, and we had our water supply! Later,

MICHAELA WELTI

we built an insulated pump-house with an electric pump for the house supply.

Our own water! Another problem solved. We felt so lucky and happy, having our own land and seeing our dream come true with each little achievement. For me, it was security and freedom, and for Max, it was finally something of his own—he never had a family life, never owned his own land. Now he had both, and could use his creativity and inventive abilities. I remembered the war, when I saw people forced to leave their homes and beautiful farms in Silesia, fleeing from the Russians, and most never saw their homes again. So much was destroyed. When we got home to Switzerland in 1945, I realized the value of a house and property, along with the value of having food, when none is available.

Max worked so hard, and as a result got skinnier all the time, though we all ate good meals and baked our own bread. But he was healthy and happy. I helped him whenever I could, holding up walls or hammering nails. One day, it was very hot, and I found some relief by working under the house in the crawl space, nailing on the spacers between the joists. It was cooler there, but not very comfortable. Then Max and I decided to head to the slough to cool off. We took off all our clothes, and sitting on the ramp, we used two big sponges to sponge ourselves off with the very cold water. Nobody could see us! It was a delicious way to cool off in the heat.

The kids were very happy at school, and little Alex played in the shade of the house, pushing dirt around with his toys. After school, the kids would help us put in a little vegetable garden. On Sundays, we would explore our surroundings, and found a huge cottonwood tree down by the slough, which the beavers were trying to fall. We wondered if such small creatures would ever get it to fall. But every year they worked at it, and finally, a few years later the big cottonwood fell into the slough. "If at first you don't succeed, try, try, again." A lesson for us all.

FROM THE GROUND UP

Sometimes, we took a picnic lunch with us and went further up the Lillooet River where the river ran wild and covered a wide area. There were wide sandbanks with pools of water that had been warmed by the sun—a perfect, safe place to swim. There was sand and clean driftwood for sand-castle building. We had some fine times there.

One day, some other neighbours down the road, also Dutch, came to visit. Luke and Doris van Beem were an elderly couple. Luke had been born in Indonesia and immigrated to Canada in his younger days. Luke had brought us a Jersey cow. Max tied her to some bushes by the driveway, and when Mike and Sonja came home from school, they were scared of her; she had such a dark face and would follow them around. Sindy was her name, and Max milked her every day. So now we had milk and cream, and tried making butter from the surplus cream. Jerseys are known to give lots of cream.

Before we built the house, doing the laundry was a problem, as I had to go all the way into Pemberton to use the laundromat. I tried doing some laundry at home, and strung a wash line between some cottonwoods to dry it. We had good weather most of the time, but in summer, the road was dusty and stirred up by the logging trucks, and often the wash got dusty while it dried. So then I had to go down to the slough and rinse it out again.

The house was coming along very quickly now. Max was nailing the sections of the walls together on the ground; then we set them up and nailed them in place. We had one short thunderstorm with a downpour, but it did no real harm. Max dug out the hole for the septic tank by hand, and put in cement walls. I said that we needed to get the kitchen and bathroom functional first, as they were most important. We had ceramic tiles for the floors, and some mosaic tiles, which came glued in squares on a mesh. We laid all three floors down: kitchen, bathroom and mud room. Tony

73

MICHAELA WELTI

recommended another Dutchman to us—Charlie Marinus. He was experienced in carpentry and electrical wiring. He started on the kitchen cabinets for us.

We soon got to know more people. The kids made their own friends at school. One morning, I was making some bread dough and there was a knock at the door. Two Mormon missionary girls were on the other side, offering to help us on the farm. I only knew of the Mormon sect from the Zane Grey books I had read in Switzerland. But the help of these religious sects always came with some hint of promoting their religion, and we were too busy to discuss the bible.

Our lives in the cabin became a bit claustrophobic, and we longed to get into the house, but if we got help from someone, it was through an exchange of work, not of religion. As school closed for the summer holidays, the kids went swimming with their friends in the slough or at Oxbow Lake, which was located across the Forestry Bridge. There was also the school farm, which had animals they could pet.

Moving Day!

Tony helped Max to nail on the metal sheets for the roof, and then, about the middle of August, we moved into the house. The hydro power and telephone were hooked up after the electrical work had been inspected and passed. Charlie Marinus had built the kitchen cabinets for us, but because of the L-shape of them, we could only get them into the house through the big kitchen window. The doors and drawers we made inside the house.

What a great day it was when we finally moved in! We were so relieved to have more room, screens on the windows to keep out the mosquitoes, proper doors, running water, electric light, and an electric stove to cook on and bake in. We could also move our

FROM THE GROUND UP

fridge and freezer into the house. Realtor Bob Taylor came by one day and could not believe we were already living in the house! All the outside walls were now insulated and some of the inside walls, also. The finishing of the inside walls and flooring could be done later. Next was to put the siding on the house, and have three sets of entrance steps built – one in the front and two in the back. For now, we had wooden ramps to get into the house.

After a while, we heard that some Swiss families were living in the valley. There were Gus and Leni Zurcher, the Stahlis, the Zurbruggs, and the Fotches. In time, we got to know them all. One day, Leni and Gus Zurcher came for a visit and introduced themselves. It was nice to speak our language with them, and we became friends with all these families in the years we lived in the Pemberton Valley.

With winter coming, there was a lot more to be done around the place in the fall. Max's brother Karl in Switzerland told us he would be coming for a visit for three weeks. We picked him up at the airport in Vancouver. The kids were delighted to get to know Uncle Karl.

Our cow, chickens, and rabbits needed a shelter. Our empty cabin got pulled with the tractor to the front of the house, to serve as a workshop, and Max planned a shelter from poles he had felled. It had enough room for the cow and a small chicken coop, with some room for the hay. Karl helped Max a lot. They were always outside! But then the fall rains started and it rained hard for two weeks. Despite this, the two brothers would put on rain gear and continue to hammer down the roof on the small barn. Karl didn't mind, although he realized what a great task it was that we had set for ourselves. This was real pioneering! Karl liked hunting, and before he returned to Switzerland, he wanted to see a bear.

We decided that I would pack a lunch and we took off for a hike up the valley. This was new for me, too—I had never been so far up the valley. The Lillooet River here covered the whole valley, spilling wild all over. We came to where North Creek runs into the Lillooet River. This is a wild river with lots of boulders. We looked and looked, but never saw a bear. There was a log bridge over North Creek, and we crossed over, but the creek got even wilder, so we ate our lunch on the banks of the Lillooet River and then turned back. We saw lots of grouse but no bears. Then suddenly, we saw a smaller bear crossing the road. So Karl got to see his bear at last!

The children liked Uncle Karl a lot, and in the evenings, when Karl would sit and light his pipe, they had fun teaching him English. When we returned him to the airport, we were singing all kinds of songs we had learned from the popular songs of the day, especially those of Peter, Paul, and Mary. We could tell Karl had fallen in love with Canada. He and Alice would migrate to BC some years later.

So far, we had put up the house, pump-house, and animal shelter, and Max had built the cement stairs leading into the house. He now started on the siding, and he also built an A-frame for our firewood for the winter, near the back of the house. Max had taken on a part-time job, splitting cedar blocks for shakes. He then got a job in logging, and so I finished up the siding and made the doors for the kitchen cabinets.

The winter of 1967-68 was mild with not too much snow, but enough for the children to start enjoying some winter sports. They hadn't experienced much snow on the coast. I had learned to ski and skate at an early age in Norway, so I knew how to teach them. I had actually brought my skis, which had metal edges, over from Switzerland. In Disentis, the snow was alpine—powdery and light. There, one could go "free skiing" (no ski lifts) in powder snow on the local hills. If we needed a ramp for jumping, we built it ourselves. The slopes for downhill we packed down by side-stepping while going up.

In the Pemberton Valley, the snow could sometimes be heavy because of its location in the Coast Mountains, where rains from the coast would turn into snow. Sometimes, the snow would turn to rain and cause flooding. In a mild spell, the rain would form large puddles that created a good ice surface for skating when it froze again. So the kids learned skating early. They formed hockey teams with their friends, and played hockey for hours outside in the fields. Even I went out on skates again, and found that one never forgets how to do such sports.

Max was not so fortunate to have been taught skating and skiing. His foster parents were too poor to afford skates or skis. The conditions for the orphan children were bad at that time in Europe, when he was young. We tried to put skates on him and gave him a chair to hold onto, but he did not feel safe on the skates. Despite this, he still played hockey with them, sliding in his boots across the ice. We had lots of fun. One winter, the Oxbow Lake over at the school farm froze over and the ice was completely clear. One could see the grasses in the water move under the ice. Everybody came to skate, young and old. It was a special unforgettable event.

Adding to the Family

In February 1968, I found out that I was pregnant again. Well, that was fine now that we were in our new house on our own property. We had achieved what we had dreamed about, and were still young and healthy. We had nothing to lose. We had gained experience and some freedom, but there was much more to learn and discover about nature.

Back in Powell River, we had started to read the bible regularly, and to compare the answers given by the Jehovah's Witness people to those in the bible. We read in Revelations 18:4 that, at the end of the world, a time would come where mankind would be very corrupt and deceitful. It was called "Babylon the Great". God would destroy Babylon within one hour. But how long would this "hour" be? And when would this happen? We did not know. It also said that believers in God should get out of this Babylon or they would partake of the plagues that will befall her.

Many prophesies in the bible are symbolic, and if one has difficulties understanding them, then only prayer to a personal god can help guide you. Then you will understand the meaning of the words. The bible reveals many signs that would occur before the "End Time." This impressed Max and me, and we had the feeling that we were experiencing the beginning of those times. Through reading the bible, we got more interested in politics, economics, and history.

In the winter, we could not do much outdoors. There was no TV yet in the valley, and no computers. So we would play games in the evening with the kids after they did their homework. Games like Monopoly and Scrabble. We got more interested in their schooling. And despite Max and I having only elementary schooling, we could keep up well with their studies and help them with their homework.

Organic Beginnings

In the spring of 1968, we found a new market for vegetables in some Vancouver stores. But first, we had to clear our meadow of brush and small cottonwoods. Our old Oliver tractor did not start so easily now, so we had to pull it with our truck to get it started. Once it was going it ran well, and with a chain on the back, Max ripped out the smaller trees and shrubs. We planted three-quarters of an acre in winter squash (Hubbard type) and one acre of rutabagas. We also increased our home garden for self-sufficiency. I

FROM THE GROUND UP

got more interested in making butter and cheese, and we started to make wine out of dandelions, blackberries, parsnips, parsley, wheat with raisins, and mountain-ash berries.

We picked our own herbal teas. Starting in May, we picked Hawthorn flowers, St. John's Wort, yarrow, wild strawberry leaves, wild roses, young birch leaves, and elderberry flowers. In the fall, we harvested a variety of berries and rose-hips. In my herbal beds in the garden, I planted bee balm, lemon balm, mint, Echinacea, sage, and meadowsweet. Later, we planted a lime tree (Linden), which has wonderfully fragrant flowers when it blooms. I blended some of the herbs for teas, which lasted us all year.

I needed knowledge to do all these things, so I joined the open library in Victoria. They sent me catalogues and lists of books free through the mail in zip bags. We read many books on farming, especially on doing things organically or according to nature's way. Especially useful were the books from Louis Bromfield, Rodale Press, Sykes, Turner, Edward Faulkner, and Sir Albert Howard. Also *The Biodynamic Method* by Rudolf Steiner. I bought several herbal books: Dr. Vogel's *The Big Pharmacy* in German, also James Duke's *The Green Pharmacy*. When I read *Of Man and Plants* by Maurice Mességué, I was very impressed, and later obtained all his books from Europe, written in German. We did not have televisions or internet at that time, so winter was a good time to read those books.

It's a Girl!

On October 27, 1968, Heidi-Anne was born in Squamish Hospital. There was no facility in Pemberton for birthing mothers. The doctor came just once a month. I found I could stay with a Swiss family in Squamish. Mrs. Waffler had been a midwife in Switzerland before coming to Canada. I was lucky to have her

79

by my side in the hospital for the birth. Heidi came very fast and everything went well. My baby was very healthy and very pretty. A "real doll!" The lady doctor who delivered her commented that if she could have a baby like that, she would not mind having ten of them! I felt very healthy and strong, too, from working outside in the fields, helping Max harvest the turnips and squash. These vegetables all got sold that fall, and that gave us some extra cash.

That winter of 1968-69 was very, very cold. It started first with a snowfall at Christmas, then the north wind really started to blow. We had some friends over for a visit, and we started to see the frost forming inside the house, around the doors and the windows. It was getting worse by the hour! We moved our chairs close to the oven to keep warm. Our visitors left early, afraid if they left any longer, their car would not start. I put mittens on the baby overnight, and Max had to get up overnight to keep the oven going. That night, the thermometer dipped to -45°F. It lasted about a week. The kids wanted to go and visit their friends who lived about two miles away, and we told them to pull their scarves over their faces or they would get frostbite from the wind. When they were late coming home, Max got worried and went down the road to get them, but found them already on their way. I never in my life experienced such cold—but it never repeated itself all the years we were in Pemberton.

After Heidi was born, our old 1952 Pontiac broke down, and we needed another car. It was winter, and we had to go to Vancouver. Our neighbour Luke van Beem loaned us a pickup truck, and we took the baby. After the cold snap, it snowed a lot, and the snowbanks on the sides of the road were so high you couldn't see over them. But we made it through and bought another Pontiac sedan, which we kept for a few years until we could afford something new. Our first new vehicle! A pickup truck!

Clearing the Stump Farm

Max had been working steadily in logging since we built the house. From the start, he had split cedar shakes from old cedar we'd found on our own property. There was a market for cedar shakes in Whistler, which was starting to develop, and had lots of building going on. He worked for Cascade Fir, Le Blanc Bros., and CRB. Then, having his blasting ticket, he worked for Squamish Mills, helping put the road through to Meager Creek. He was later employed in blasting again when they built the Duffy Lake road through to Lillooet from Mount Curry near Pemberton. Apart from his job, he was always busy improving the land. Whenever we could afford it, he hired a Cat driver to clear more land of tree stumps, slowly making fertile land from our "stump farm." There was a young Dutch couple who had a farm a few miles down the road—the van Gouls. When Harry van Goul got a better job that took him away from Pemberton, we rented their farm. He had a root house on the place, and enough cleared land to grow potatoes and other crops. It was there that we planted our first potatoes and oats.

Planting seed potatoes.
Photo credit: Welti Family photos.

We planned to build our own potato storage house just across the road from our property. Eventually, we wanted to get into growing seed potatoes. The Pemberton Valley is known for the quality of its seed potatoes, most of which are sold to the US. Seed potatoes come under the BC Certified Seed Growers Association, which is government controlled, and does regular inspections. But each farmer has to look for their own market. The potatoes are graded and sold again for seed production.

In 1970, we had visitors around Christmas time from Australia—my brother Jacki and his wife Mira. Also my eldest sister Carmen. Then in 1971, my mother came in November for a visit, which lasted a few months. She was seventy-six years old and made the long airplane trip all by herself, without knowing much English. I had not seen her since leaving Switzerland thirteen years before. When Max and I picked her up at the airport in Vancouver I was overjoyed. She had seen so many countries in her life, and I was astonished and full of admiration at her courage in tackling such a long trip by herself, at her age. She enjoyed the children so much, especially little Heidi. She stayed with us through the winter, helping me harvest carrots and babysitting. It was so good for the kids to enjoy their grandmother.

Again around Christmas, we had another visit from Jacki and Mira, with their son Peter, all the way from Australia. It was a mild winter with not much snow. Now we had more animals around the farm, including several dogs: a black Labrador and two German Shepherds. Besides the milk cow, there were two geese, several ducks, a pig, and a few Hereford heifers.

I learned how to make cheese and butter. For the butter making, I skimmed the cream from the milk every morning and ladled it into an ice cream pail, which I then put in the freezer. Each day's cream was added to the pail and, when the pail was full, I thawed out the cream and put it into a two-gallon plastic pail with a lid,

FROM THE GROUND UP

into which I had cut a hole in the middle. We bought an electric beater with a small motor that fit into the hole. All I had to do was plug it in and let it run until the cream got clumpy. Then I gathered the clotted cream together and the washing of the butter began. The washing took longer than the churning of the cream with the machine. I poured the churned cream through a sieve and gathered the clotted clumps together, then started washing the butter with cold water. Then I turned it onto a slanted wooden board, where I kneaded it to press out the remaining water. This yielded sweet butter without any salt. Then I froze the butter in tubs in the freezer.

Making Cheese the Simple Way

I had bought a book on cheese-making, but our Swiss friend Gus Zurcher taught me a much simpler way. In his younger days, he had worked with dairy cows in the Swiss Alps. In the summer, they would drive the cows higher up into the alpine meadows, where they had aromatic herbs and grasses to eat. They had no electricity there, but the cows had to be milked every day, and something had to be done with the milk, so they made cheese from it with a simple method.

I started with about ten litres of milk, either whole or partly skimmed. If you have a Jersey cow, which gives a lot of cream, you can skim off most of the cream for butter, and make cheese from the low-fat milk. I used an enamel pot that I used also for canning, or a stainless-steel stock pot. You will need rennet to coagulate the milk, and a dairy thermometer. The milk gets heated slowly until it is 95-102°F. Dissolve the rennet tablets or powder in water according to the amount of milk, then pour this slowly into the milk, stirring with a wooden spoon. Let it stand with the lid on for fifteen to twenty minutes.

MICHAELA WELTI

The milk has now become jelly-like, as the protein coagulates, and so has to be cut with a long knife that reaches to the bottom of the pot. Starting on one side, make parallel cuts about an inch apart. Then turn the pot and make crosswise cuts in a checkerboard pattern. Now the milk has to be heated up again, slowly, and the curds cut again using a wire whisk. Draw the whisk carefully through the curds until the curds get smaller. You can also use your fingers to break them apart. The heating of the curds makes them drier, and you keep working it until the curds are about the size of rice grains. Shut off the heat and let the curds settle to the bottom of the pot—about ten to fifteen minutes.

Now comes the tricky part. Roll up your sleeves, and go with clean hands into the pot without stirring up the curds. Lay your hands on the curds and press down carefully a little bit, gathering the curds into a ball with your fingers so that they're formed into a loaf. The curds will stick together so that you can lift the loaf out of the whey with your hands. If you have a lot of milk, say twenty litres, you will have to use a cheesecloth under the gathered loaf to lift it out of the whey.

The mold you use should be round, preferably made of wood, with drainage holes in the bottom. Put the loaf into the mold and let it drain for twenty minutes. You do not have to press the cheese; it will press itself with its own weight. Turn it every half hour at first, then every hour, and, later, two or three times that day. Then let it drain overnight. The next day, make a salt solution, dissolving the salt in boiling water. Pickling salt is best. Cool the salt solution before adding it to the cheese. Use sufficient salt so that the cheese starts to "swim," lifting off the bottom of the bowl. Let the cheese stay in the solution a few days. The salt will slowly permeate the cheese and also make the outside rind slimy.

Now smear the rind smooth with your hands before putting it on a wooden board to dry. When the rind is dry, wrap the cheese

84

FROM THE GROUND UP

in cheesecloth that's been soaked in brine then wrung out so it is just damp. I usually put the loaf of cheese in the fridge to cure, turning it occasionally, for about two months. I leave the brine-soaked cloth on it to prevent mould growth. The cow whose milk you use for cheese-making should be healthy. If she has mastitis (an infection of the udder), don't use her milk. In the spring when the grass is first growing, don't make cheese from cows grazing on it. And the cheese should not be made from cows eating fermented hay or silage, as this will make the cheese bloat or crack.

I compare cheese-making to wine-making: cheese has to ferment and cure just like wine does. Sometimes, there was a difference in the flavour of the cheeses. This mostly depends on what the cows were eating, or variations in the making of it. But it just made for interesting variations in each batch. A cheese can be eaten at any time, from fresh until aged, with wine or bread. The cow that did best for us was a cross between a Hereford and Jersey, and she was quite selective in what she ate. When Max fed her oat-hay (oats cut while still green) and carrots, she gave the best milk for cheese.

Self-Sufficiency

In time, we also raised two pigs. Max built an enclosure for them from logs, under the open pole barn we had first built by the house. When our old mother hen had another flock of chicks, she would range free with them around the barn. The chicks liked to hop on the logs of the pigpen and play around in there. I started to wonder why the chicks seemed to be slowly diminishing—every evening when I closed them in, there were fewer chicks! I guessed that the pigs ate them, one by one. But I still had some left.

We would also do our own butchering. We had help from an older man who showed us how it should be done. It is more complicated with pigs, since they require soaking in a trough of very hot water to soften the bristles, which are then scraped off. Gus Zurcher also taught us how to brine the bacon and ham the Swiss way: in a brine prepared with salt and spices such as celery, bay leaves, Juniper berries, peppercorns, rosemary, sage, and thyme. You boil this up and let it cool, then let your bacon or ham pieces soak in it for about a week or longer. If the pieces are thick, they will take longer.

Max invented a smoker. He put up some poles, four of them, ten feet high, with a crosspiece on top from which to hang the pieces of meat. You could raise or lower it from a pulley-block. He wrapped this structure in a plastic tarp, leaving the top and bottom open. Then he built a fire of alder wood on the ground inside this tent. It's important that the smoke has cooled when it reaches the pieces of meat. In Switzerland in the olden days, the farmers would hang their sausage, ham, bacon, and jerky in the chimney to cure. Those big farmhouses had big chimneys.

We made some jerky, too. For this, you use only really lean meat from a flank of beef. It was also soaked in brine solution with added spices, then air-dried. Canton Graubünden, where I come from, was famous for its *Bündner Fleisch*. It was so delicately made and air-dried that it could be shaved into very thin slices, which curled up into rolls. It is very expensive to buy now in Switzerland. They say the dry mountain air contributes to its delicacy. Our jerky was never quite as good as that, but it was still a big hit with the kids! When they brought friends over, they could never get enough of it. We also made sausages. I ground up herbs like rosemary, sage, and thyme into the meat, along with lemon juice to give the sausage meat a good colour, instead of saltpeter.

Hippie Invasion

About the time that Max built our first root house, we started our first market garden. We planted lots of different vegetables—corn, carrots, cucumbers, beans, cabbage, cauliflower, beets, etc. It was a small business based on a "pick-your-own" model. In time, lots of people came. Too many, in fact! Max was working and the kids went to school in the fall. Only I and little Heidi were at home. When the summer came, Sonja, my eldest daughter, could help with the people wanting to buy our vegetables. They would even come from Whistler. We got to know all kinds of new people, even the hippies who were arriving in the Pemberton Valley at that time. I was interested in their back-to-nature cause. They camped by the river and ran around naked, trying to plant gardens, keeping a few goats. But their ideas were so bizarre, and they had no knowledge of how to survive. They lived mainly outdoors. The mosquitoes were bad, and in winter, there was the cold and snow, which drove them out eventually. The farmers did not like them.

One day around that time, a truck full of hippies arrived outside the house. I looked outside and saw some very nice-looking girls with long, curly hair and colourful dresses. But then I watched them leave again without saying anything. *That's strange!* I thought. Then later, when I opened the door, there was a person sitting on the steps! He must have been left behind by the truck. I saw that he was kind of a cripple, a young man with red hair and beard. His name was Bruce Stephenson. I took him into the house and we had an interesting conversation. He was from Vancouver and quite educated. I did not want to turn him out with no car and disabled. What would he do? Where would he go? For a while, a bit of a friendship developed, but my family did not like him too much. He tried living "in the bush," across the river in a lean-to, with a goat he milked when he needed milk. He planted some carrots under the cedar trees. When the winter came, he left. But I saw

MICHAELA WELTI

him later, at a logging festival. He did not recognize me. His face was very pale, and I knew he was taking drugs.

No Latch-Key Children for Us

During and after World War II, many women went out working. People in Europe were eager to rebuild, and since the men were at war or had been killed, the women took their place at many jobs. And they continued to go to work after the war. I thought that maybe I should get a job, too. We thought this over very carefully. We would have to buy another car, more groceries, and more clothes. The nearest work was in Pemberton, which was sixteen miles away from home. It was better to keep things as they were! I could contribute to our income by growing our own food, saving on groceries, making butter and cheese, baking bread, sewing, and planting the market garden.

The younger generation who came from the post-war world seemed to be turning their backs on their parents' way of life. In Europe, they called these children "latchkey children." When they came home after school, there was nobody home, and so they had the key to their house or apartment on a string around their necks. The hippie movement evolved because the young people were disillusioned with what their parents had done with their world. This back-to-nature revolution was a new historical event.

We kept our market garden for about three years, and it was so popular that we were sometimes overwhelmed with people. Even on Sunday when we were closed, with a sign out on the road, they would still come. I fell behind with my housework and had to leave Heidi in the house while I served customers. But she was just a little toddler and often got into trouble. So, instead of the market garden we seeded about two acres in rutabagas, and gradually built up a "quota" for them. A quota was necessary to sell your crop to the marketing board.

Clearing More Land with a Cat

Max had built his first root house. It was thirty by forty feet, and this allowed him to store quantities of potatoes, carrots, and rutabagas. Meanwhile, again hiring a Cat, he cleared some more patches of land. At this time, he was working for Squamish Mills. His boss there was Pat Brennan, who was also mayor of Squamish, and liked Max a lot. He saw that Max had really struggled to get his farm going. He had seen, too, that we knew how to grow quality produce from our success with our market garden. Now he arranged for a Cat and a driver to clear land for us for a whole week for free! This was a great help. It would also enable us to apply for a long-term farm loan. One had to have a certain acreage cleared before applying for such a loan.

We went out into the field and watched the Cat working at the clearing. When he pushed those giant cottonwoods over, complete with roots, water would well up in the hole left behind, as if a spring was there, but it would soon subside. The cottonwoods suck up a lot of water for themselves from the earth. After the clearing, there were lots of broken branches and roots left behind. The Cat had pushed several heaps of stumps together into huge windrows for burning later, and what was left had to be picked up by hand. We went out with the kids and picked up the branches and roots and threw them into a hay wagon with sides on it that Max had rigged up. When it was full, we dumped the wood debris into piles at the edge of the field.

It was slow, back-breaking work. So Max invented a rake: he fastened a long beam to the three-point hitch on the tractor, so he could raise or lower the beam. At the end of the beam he fastened a chain harrow. This way, he could rake the waste wood into rows, for burning later.

Then came another dilemma. When we walked in the cleared fields, the soil was like moon dust. So powdery. It was like water flowing around our boots, and very dusty. It didn't seem to have any structure. Where had all the topsoil gone? The blade of the Cat had pushed it into the stump and slash piles that were destined for burning! Later, Max tried seeding some grain into this soil, but it only grew a few inches and then dried out. Max was bitterly disappointed about that. But now we had enough land cleared to get the long-term farm loan, and this enabled him to buy more tractors and other machines.

We learned later that the Cat that cleared the land should have had a clearing blade, a huge rake that will pick up the stumps and trees and leave the topsoil behind on the land. The land we had cleared had mostly cottonwoods and birch on it. In the past, it had had big stands of cedar trees that were logged off in the 1800s, so there were many huge stumps left behind, some ten or twelve feet in diameter. After removing these stumps, we could not expect to have good soil, but this realization became a lesson, which I will mention later.

Maturing Family and Farm

In 1974, we were able to go and visit Switzerland and our relatives. It had been such a long time since we had seen our families. We took little Heidi with us, who was six years old. Sonja (16), Mike (14), and Alex (11) stayed home on the farm. It was July and the kids were on holiday. They were now quite able to look after themselves and the farm. Usually, when they came home from school, we would gather around the table and talk about school and the day's events. Max would take the boys outside and teach them how

FROM THE GROUND UP

to drive the tractor, split firewood, shovel potatoes into the grader, and lift and stack 100-pound sacks of potatoes or turnips.

In their free time, they could visit their nearby friends. We were sixteen miles up the valley from Pemberton, so the kids could not easily run into town. Max gave them responsibility and trusted them early on to do a job by themselves. They could build a fort in the woods if they wanted. When we went to Switzerland, Mike did the haying and Sonja looked after our beehives. I taught the girls cooking and baking from scratch using basic ingredients. We bought a piano again, and Sonja had piano lessons. Both girls had horses later, and the boys had dirt bikes. Alex got a snowmobile. Cars came later, when they were sixteen years old.

When Heidi was born, we fitted out the attic in the house, so the boys could have their own room. Later, when they were older, Max let them grow their own crop of turnips, and when these were sold, the boys shared the money. They also made some money doing custom work with the tractors. Max got more into the seed potatoes, and we also grew carrot and turnip crops. Max built a big pole barn for our hay crop, which would feed the cattle through the winter. A second root house got built, this one measuring eighty by forty feet.

There was a Dr. Dill who had come to the valley before us. He had bought up some nice pieces of land near Pemberton in the upper valley past our place. He was living in the Philippines and had a hospital there. He was not an active farmer, but had some buildings and machinery on his land, and a manager who came sometimes to work it. He was also interested in organic agriculture. Max decided to rent the upper valley farm from him to grow potatoes, for a lease of $1,000 a year. Max also built some fences, and had cattle up there. Now, in 1976, we finally had enough land that we could farm full-time.

91

In addition, we bought forty acres of land adjacent to our property. Max worked this land up and planted some crops there. Later, he sold the forty acres to Mike and Alex for them to farm. When we started to have good results growing and selling carrot and turnip crops, neighbouring farmers also started growing these, especially turnips. They got sold to the marketing board in Vancouver, and together with our neighbours, we were able to build up a nice quota system for turnips. Max was also the first to build a bunker silo in the pole barn. This came in handy for feeding cattle in winter when the hay crop was not so good.

The seed potatoes got inspected for viruses in the fields during the growing season. The crops also got sprayed for different diseases, the turnips and carrots for worms. We also used artificial fertilizer. But we were trying more and more to eliminate all this spraying, something with which we succeeded in our own garden, with good results. One year, I seeded some wheat in our garden to try out green manuring. I had seeded the wheat quite densely, and it came up a nice green stand. When it was about ten inches high, Max turned the sod in, and the next year the soil was loose: nice and crumbly. We found now that the cabbage and turnips had a better flavour when cooked, without the intense flavours they sometimes had before. The broccoli, carrots, and even potatoes had better flavours. Yes, we had pests—worms and aphids—but now there were new organic sprays on the market like bacterial Thuricide, which we used. We learned a lot from the many books we read on organic agriculture. Max became more and more convinced that it is the soil that must be healthy and fertile first to solve the problems of pests and diseases in plants, animals, and man.

With the land we rented from Dr. Dill on which we grew our crops, we could have farmed very well and made a profit—but it was not our land. Max would have worked the land up for somebody else, with no guarantee that it would be his in the future.

FROM THE GROUND UP

Our garden served us an example of organic methods and how they worked out. The less we sprayed and avoided chemical fertilizers, the less we saw pests and disease infecting our vegetables. We had a good market for turnips, carrots, and parsnips, which we sold from the farm in the fall. But neighbouring farmers also started to grow these crops. In February of 1983, I returned from a trip to Switzerland to find it to be a bad year for selling seed potatoes with very low prices. Max was distraught, and decided to install a cooling system in the small root house. If you could keep the potatoes until June or July, you could get a good price for them, since the restaurants always wanted them for French fries or potato chips. The first year we had the cooler, Max made $20,000 from the potatoes he kept this way, selling directly to the restaurants. But it was hard work, loading and unloading the 100-pound sacks into and out of the cooler, and time-consuming to haul the potatoes to market. The cooling system also benefited our other crops of turnips, carrots, and parsnips. We were the first in the valley to build a cooler, but the other farmers gradually copied this.

*

One year, our eldest son Mike stayed home from school to help Max on the farm, with the intention of seeing if he would like farming for a living and might take over the farm eventually. Mike was very businesslike and a good talker. Max even paid him for working on the farm. But Mike wanted to run the farm like an eight-hour-a-day business, which one cannot, with a farm. He became a very successful salesman for Snap-On Tools, with a high income, who lives near Vancouver.

Our other son, Alex, was interested in mechanics, and when he was in his teens, he got a job on the weekends as a carpentry helper, building with a young Swiss man. After his graduation, he went

93

MICHAELA WELTI

into an apprenticeship with an electrical contractor. The Whistler ski resort was booming by then, with lots of construction and many jobs available. Both boys were very good at drafting at school. Alex stayed for a while in the valley where he built, by himself, a big garage/workshop with living quarters on the forty acres we sold to him and Mike. Max loaned him the use of his tractor and tools for the building. He finished his apprenticeship in Vancouver and is now a licensed electrical contractor and renovator.

A Different Kind of Cat

I remember a time when Alex was about ten years old. I was waiting for them to come home on the school bus. Suddenly, there was Alex, running as fast as he could, bursting into the house, yelling, "Where's my BB gun? My BB gun!"

I said, "What's happening?"

"A cat!" he said, "A cat up in the tree!" And he got the gun and ran out. I followed him. Surely he would not shoot a domestic cat? I heard his shot, and when I came to the bus stop, all the kids were standing in a group looking at a wild bobcat, lying dead on the ground. His shot had gone right through its eye into the brain, and the cat fell out of the tree, quite dead from Alex's amazing lucky shot!

Well, we took it home, with Alex very proud. His lucky shot had not damaged the beautiful fur of the animal one bit, so, since I had experience skinning rabbits, I skinned the cat and tanned the hide myself. I thought about what I would do with that lovely fur. In the fall of 1976, I was to go to Switzerland to visit my family, and I had sewn myself a new winter coat. I thought that fur would make a lovely collar for the coat. A real attraction. Well, it did, but not the kind I was expecting! In Zurich, we had to get on a bus to go to the airport. People were smoking and reading their newspapers, but I noticed some funny looks from people, which they hid behind their newspapers. It was because of my wild fur collar! I didn't realize that the anti-fur movement had made it unfashionable to wear such things.

Bobcat painting by Yours Truly.
Artwork by Michaela Welti.

Weather Issues

We experienced flooding three times in the years we lived in Pemberton Meadows. One time, the flooding came at the end of August when the potatoes were still in the ground. Many farmers lost their entire crop of potatoes. The flood I remembered most

MICHAELA WELTI

came after it rained for many days without any let-up. I went to Pemberton to do some shopping in the afternoon. When I returned, I saw that Ryan Creek, which comes out of one of the side valleys and runs parallel to the road, was dangerously high. The water was just about to the edge of the road and made whirlpools. That night, before I went to bed, I could hear bumping noises. I opened the window and realized what I was hearing was stones and boulders tumbling down Ryan Creek. I woke up Max, and we listened for a while.

When we looked out the next morning, everything seemed normal, but Alex came by quickly and told us to check out the forestry road. Everything was under water at the van Loon farm, except the house and barn. The cattle had gathered together on the high ground, and they had to get them out using a boat, driving them through the shallower places to bring them home. We went further down the road and saw many mice and small animals running down it, rescuing themselves from the floodwaters. We saw hay bales floating away down the river of water. Pemberton was even worse. The village had an elevation of 700 feet. So about a hundred feet lower than our farm. The floodwater was six feet high at the high school. The water gradually subsided, but there was a lot of damage. This happened about 1980-81.

The older farmers told us of a worse flood in 1948, when the water had been so high that people had to be rescued with boats, or had to swim. They remembered their chickens rescuing themselves by climbing onto floating pumpkins. What a strange, grotesque sight, to see your chickens floating away on these pumpkins! Our son Mike was visiting a friend in Pemberton, but could not come home for two or three days because the Ryan Creek Bridge was damaged.

We also experienced heavy snowfalls in winter. It could snow for days. Sometimes, it was dark in the house, as the snow sliding off the roof met the snow piled up under the eaves. The dogs would climb up on the piles outside the windows and "smile" at us, sitting high on the snowbanks. One time, the snow reached up to the kitchen window, which was six feet off the ground. Max made a mark on the telephone pole to record the height. The following spring, the snow was still there at the beginning of May. Then it got warm enough that you could actually hear the snow and ice melt. Little Heidi had her bathing suit on, playing in the melted puddles. Me, I was so sad because I could not get my garden started.

Seed potatoes growing in field.
Photo credit: Welti Family photos.

Like the other farmers, we were getting more into growing quality seed potatoes. I would like to explain the system that was required: the potatoes are grown under the British Columbia Certified Seed Potato Growers Association, which is partially funded by the BC government. They are inspected several times in the growing season. The certification program is now the responsibility of the Canadian Food Inspection Agency. There is an established Elite system: Elite 1, 2, 3, and 4. What this means is that the inspectors take several potatoes of different varieties from different fields, and send them to the University of British

Columbia, where they are inspected in the university laboratories for viruses. The virus-free ones get distributed to farmers who have a greenhouse. We had four such greenhouses in the valley. When spring comes, the potatoes sprout, and these sprouts get cut off and transplanted in the greenhouse. The sprouts grow roots and become plants. To increase the number of plants, stem cuttings can be taken again, two or three times from each plant, and planted in sterile soil to make roots. These plants get distributed to several farmers and planted in their fields. The resulting potatoes are called Pre-Elite. The following year, these Pre-Elites are planted to produce potatoes. These are called Elite 1. Elite 1 produces Elite 2, which produces Elite 3. Every year, the stem cuttings are started again. The farmers had to work together on this to upgrade their potatoes.

Later, in the 1970s, we got our own laboratory in the valley, and things got a lot more complicated. Everything had to be sterilized. Every farmer had to do their own seed potato upgrades. We had a pressure cooker to sterilize the equipment used for tissue culture. We had to remove our shoes when we entered the building. We worked on tables with bright lights and surgical instruments. Tiny bits of tissue are cut from the eye of a virus-free potato and planted into sterile agar-agar in Petri dishes. When they sprout and are one to two inches high, they are cut again and transplanted. When we handled these tissues, we used alcohol to sterilize the instruments, knives, and tweezers and then flamed them before use. The results of growing them in screened greenhouses were small, round seed potatoes. The screens were necessary to keep out insects, which can carry viruses. In this way, we got virus-free potatoes of the highest quality.

FROM THE GROUND UP

Our Indigenous Neighbours

For seeding and planting we hired a crew. They were mostly Indigenous. The Lil'wat or Mount Currie First Nation was near Pemberton, one of the largest in BC. Sometimes, they would come out to the farm in their own vehicles, or we would go and get them. We were lucky to have the same family, the Sams, year after year. They did not drink alcohol. We had to provide lunch for them, and a coffee break in the afternoon, which I brought out into the fields. Over the years, we got to know them better and learned a little of their lifestyle. Some of them were quite educated and had gone to university, like the Williams family. They had cattle and a lot of land.

Sometimes, Max would have to go early in the morning to the reserve to collect our workers. Lots of them would be drunk, especially after a weekend, and their kids would be running around half-naked, not knowing where their parents were. The ones who did not drink had a hard time trying to grow a garden or raise chickens, because the others would steal their chickens and produce, or they would cut fences to let the cattle out.

But the older people made beautiful baskets, woven from split cedar roots. Some of the older ladies came out to our farm to search for cedar roots for their baskets. They also knew a lot about herbs. At one time, I taught some art and spinning with a drop spindle at the elementary school. The Indigenous children showed the most talent for art. Aboriginals on the reserve got lots of money from the government to build houses, and some beautiful mansions got built on their land. But they did not look after the houses very well.

In 1983-84, Max decided to sell our cattle and develop the soil on our cleared land. We now had about seventy acres cleared. In the fall of 1983, we went to visit my brother Jacki in Sydney, Australia. Together with him and his wife, we travelled to Queensland and visited Brisbane. It was a lovely three-week holiday. We came home

MICHAELA WELTI

before Christmas. The next spring, Max let the pastures grow until about June or July. He now had two good tractors—a sixty and a seventy-five horsepower—and a really heavy disk harrow, twelve feet wide with a four-inch tubing frame.

Green Manure to Build Up the Soil

As I mentioned before, we had read many books on organic farming and building up the soil. The best ones, which had impressed Max the most, were two by Edward H. Faulkner: *The Plowman's Folly* and, later, *Soil Development*. Faulkner was an agriculturist employed by the Department of Agriculture in the US. He had studied the problem of soil depletion, which gave rise to the dust-bowl conditions on the US prairies in the 1930s. He believed that plowing the soil would disturb the microbes and beneficial fungi occurring in a naturally decomposing layer in the soil. Faulkner wrote: "The fact is that no-one has ever advanced a scientific reason for plowing."

By disking in green manures into the soil, he believed it was possible to turn inferior soils into high-yield croplands. *The Plowman's Folly*, published in 1943, was revolutionary and became very famous after World War II. It was called "One of the most revolutionary ideas in agriculture history." With the money Faulkner made on his book, he bought a piece of farmland. The soil was depleted and not very fertile. There was some old manure on the farm, but instead of using it on the land, he sold it or gave it away. He did not use any artificial fertilizers or insect sprays. He wanted to prove that repeatedly turning grasses or grains into "green manure" would build up the soil. And so it did.

Max standing in the rye.
Photo credit: Welti Family photos.

Max wanted to give it a try! Over time, Max had observed that the grass that grew naturally by the fences, where it wasn't disturbed or eaten by animals, was lush and grew better than in the pastures. So he let our pastures grow until the grasses were ready for haying. Then, instead of haying, he disked it in with the heavy disk. He left the disked fields for a week, to let the grasses break down in the soil. Then he cross-disked the fields. We had our own oat, rye, and buckwheat seed, but some of the grass seeds we had to buy. One variety of grass seed, orchard grass, was especially good. It grew very fast and developed a good root system.

After Max had disked the fields again, he would seed them with grass seed or mixtures like oats and buckwheat. Whatever came up from the old pasture was good, too. When the seeds sprouted, he let them grow to a full crop and then disked the whole crop in again. Some of the oats and buckwheat would sprout again that same season. After disking again, he would seed fall rye. This is a perennial, and will survive the winter to start growing again in spring to produce a full crop. Some of the rye he would harvest for seed. That field got disked again and seeded down to grasses, clover, oats, or buckwheat, which were single-season crops that got turned into the soil again. Seeding into the disked stubble of organic grass roots to form a new crop benefited the soil very much. This way, a wonderful layer of organic matter would develop, and every time he would seed down into a new crop of green manure, the crop would get bigger. Some of the rye and buckwheat grew to shoulder height. Max would work alternate sections of the farm, slowly building up the soil.

Dr. Dill liked Max, and what he was doing with the land. He rented us his whole upper valley farm, about 150 acres, still for $1,000 a year. Max had to build some fences there. Some fields were standing in canary grass, and he disked these up and commenced building this soil up also. We never had any cattle again. Max liked vegetable crops and wanted to increase the yield of potatoes, turnips, and carrots.

The results were phenomenal. The potatoes came up from the earth, sliding on the conveyor of the harvester so clean and healthy because the soil was so crumbly, with excellent tilth. There was such a crop of potatoes, we had to go slowly not to miss any. Everybody on the crew was excited, especially the Indigenous members. Max's success was the talk of the valley. The grasses in the hayfields were waist high, and cutting such a rich crop also went slowly; there was so much grass. It seemed like a miracle!

We observed other farmers working their fields. They plowed and disked and cross-disked, finally using the rototiller to make a smooth seed bed. But when it rained, there were sections that had standing water in the fields. Too much working of the fields with heavy equipment had compacted the soil, especially the rototiller. Max's method of working his fields by slowly building up the soil was only what nature did in the forest and wild prairie. Every year, tons of leaves and grasses die and decompose, sustaining the living trees, brush, and wild grasses. Max was just speeding up this natural process.

I would like to mention also that cattle, horses, and sheep grazing in a field will compact the soil. A cow weighs between 800 and 1,000 pounds. When there is too much compaction, it depletes the soil of oxygen, which is necessary for microbial breakdown. If Max found that the earth needed to be loosened, he would run a cultivator through the fields. The fields where the soil was poorer, like after initial clearing, he would work longer, repeating the tilling-in with green manures.

Cows in pasture.
Photo credit: Welti Family photos.

Max had a discussion with one of the other farmers about the importance of manure that cattle leave in the pastures. Max told him, "You can't tell me that what comes out of the animals is of more value than what they eat!"

We knew now that our method could be used on any kind of soil, anywhere, even in a small garden. Just remember to leave one-quarter or a third of your garden or acreage for the creation of topsoil, and rotate this every year. Grains, grasses, buckwheat, and clover are all good green manures, containing nitrogen and many other minerals. One only has to look at nature: in uncultivated fields and forests, you find a great diversity of plants, fruits, berries, and mushrooms growing in profusion without any plowing, seeding or spraying—in short, without help from mankind.

Fukuoka's Way

The Japanese author Fukuoka, who wrote the book *One Straw Revolution*, tells the story about the farm he inherited from his father, which was an orange orchard plus a rice acreage. Fukuoka was a professional agriculturist. Instead of following the traditional method of farming, with lots of pruning, artificial fertilizer, and tilling, he sought a more natural way, but was unsure about how to go about it. So he took a year off, travelling to study the methods of different cultures, watching people work their farms while meditating and pondering. When he returned to the farm, he had come to the conclusion that the best thing to do was—nothing! He left off pruning the orange trees and seeded the rice into the fields without tilling. This was a real lesson.

The orange trees responded with lots of blossoms and fruit. The rice also did well and was harvested. After the rice was winnowed, the leftover straw was returned to the fields. Again, he broadcast the rice into the fields the following season. But now he found that the birds got most of the rice seed. So he made a thin mud from clay and water, covered the unplanted rice seed with this, and left it to dry. When these coated seeds were broadcast, the birds could not find them! Always returning the straw to the fields, he built up

FROM THE GROUND UP

in time a spongy, rich topsoil. Later, he eliminated the flooding of the fields with water, which was the usual method of rice planting. Every year, his crop of rice had greater yield, eventually producing as much as, or even more than, farmers using conventional methods, and with better quality.

The saying, "Be good to the earth, and the earth will be good to you," has its realization here. Consider the importance of the microbes in the soil and the earthworms, which help break the humus down. They are better left as undisturbed as possible. We are only starting to appreciate the wondrous ways of nature. And who created all this wondrous nature? Jesus said, in the bible (Luke 12:27), "Consider the lilies of the field, how they grow. They neither toil nor spin. Yet I say unto you, not even Solomon in all his glory was clothed like one of these. But if God so clothes the grasses of the field, which is alive today and tomorrow is thrown into the oven, how much more will he clothe you—ye of little faith!" And again in Romans 1:20: "Ever since the creation of the world. His eternal power and divine nature, invisible though they are, have been seen and understood through the things He has made."

A Swiss friend of mine gave me a book by a German author named Henning. She said, "Read this! It will be very interesting and keep you in suspense!" The author was in his ninetieth year when he wrote that book. He must have been a biologist. The title of the book was *The Secret Life of Plants*. In it, he described the microbes and fungi in the soil and among the roots of plants. How they communicate and interact with each other, and how photosynthesis is achieved. All scientifically based. It was so interesting, and a real eye-opener. It all relates back to those quotes from the Bible. On the last page of his book, the author stated that after all his scientific inquiry, he could only say that there is a God.

Max now had experienced for himself the success of creating a fertile soil by organic methods. The farm was now ready to be expanded into full operation. But now we had a new experience. A not so good one! The turnips or rutabagas were a fall crop and had to be sprayed with Dazomet; a new spray that was stronger than Diazinon. The spray kills the fly, which lays its eggs on the root, and when they hatch, the worms destroy the root and cripple the turnip. They recommended spraying in the evenings and not going into the field for two days. When he went into the fields, he saw hundreds of earthworms lying between the rows, dead! They had come out of the ground to escape the spray, and had died there. It was a big shock for him, and he did not tell me for a long time.

Max was the first to grow turnip, carrot, and parsnip crops in the valley, using his root house to keep them through the winter, and the cooler in the summer. He had built up a quota for the turnips. Other farmers followed up on his methods, and wanted to be in on the quota system, too. Competitiveness and some jealousy developed.

Time for a Change of Pace

Max grew weary, and neither of our boys wanted to take over the farm. So around Christmas in 1991, we decided to try and sell it. It was a mild winter, and in the New Year ,we went for a trip over the new Duffy Lake Road, which Max had worked on as a blaster. We wanted to take a look at the Kamloops and Vernon area, to see if we might like to live there. I liked the sunshine, and somehow the open country appealed to us, after having lived in a valley closed in by mountains. When we came back, we first went to Whistler Village, where there were several real estate agencies. In the end, we decided to list the farm with a lady who lived in Pemberton. The listing price was $350,000. Max also told Mark Kurne, with whom we shared harvesting equipment for potatoes, that our farm was for sale. After a few days, he came back and said he would be

FROM THE GROUND UP

interested in buying it. Mark also had a small farm, about thirty acres, close to the river, with a small log house. He also rented a lot of land on which he grew seed potatoes. He had to sell his place to be able to buy our farm, but within a few months, everything was settled, the farm agencies came and evaluated everything, and on March 30, 1992, the sale was completed. Mark also bought some of Max's equipment. Now we had to find a place to live!

We had the option of staying in Pemberton and could have bought a smaller place with a house. Max was now in his sixty-seventh year and ready to take life a bit easier. Besides, he did not want to see what would happen to his farm, which he had created from the stump farm we bought. And we were too adventurous! We never wanted to stay in the same place! And the kids were gone! We were also surprised by how quickly the farm sold. It was as if it was meant to be. I started to clean the house and pack.

*

Max and Mark had to go to a meeting in Vernon and I was at home alone. I had a big catalogue from the real estate agents listing a lot of properties and houses all over BC, and I sat down to study it. It took me three hours until I found an area I liked. It was in the southern interior. Properties were cheaper there and the climate was milder and had lots of sun. We had looked at some properties around Kamloops, Vernon, and Lumby, but then we decided to go over the Monashee Pass to see what was for sale in the Arrow Lakes area.

The real estate lady with whom we had listed our farm knew people in the real estate agency in Nakusp. So we went and looked at properties in Fauquier, and on the other side of the lake, along Needles Road. I liked the Needles property, which had a wide lake frontage with a wonderful view, but there was a water problem

MICHAELA WELTI

there, and not enough land for Max to work. There was also a property in the Inonoaklin Valley, near Edgewood. It comprised a lovely modern log house on twenty-three-and-a-half acres, with a barn and other buildings. (I'd adored log houses since I was a girl in Switzerland, and had always secretly wanted one.) There was also a woodlot of four to five acres, and a creek that ran through the property. Around the house were western larch, spruce, cedar, and birches. The log house was not quite finished inside. It had big windows, and a veranda facing south, exposed to lots of sunshine. The two fields on each side of the creek had been used as pastures—and it was just what Max wanted.

An extra building of sixty by twenty-six had been an apple juice factory with an apple press. There was a small apple orchard behind the building. This small farm was just perfect for us. Despite the fact that we were seniors, we were still active. Max wanted to do some farming, and I wanted a big flower garden. We bought the place for $92,000. For us, it was cheap compared to land in the Pemberton area. We had enough money left to finish the log house and improve the place. Max had his two tractors and the disking and haying equipment moved from Pemberton on two trailer trucks. The furniture came in a moving van. We moved in on May 14, 1992.

4. WILD ANIMAL STORIES— WOLVES, DEER, BEAR, AND GRIZZLIES

Close Encounter with Wolves

Sometime around the end of the 1980s, in February when we were still in Pemberton, there was a beautiful moonlit night when I got up to use the bathroom. We had a dog called Rex, a Shepherd/ Malamute cross who lived outside. This night, he started to bark alarmingly. I had gone back to bed, but the barking got more furious, punctuated with whining, as if he was afraid of something. I went to the back room where there was a window beside the door, so I could see something. There were maybe five dark shapes of animals surrounding the A-frame woodshed, which was close to the back door. Rex was hiding in the woodshed. One of the animals was very close to the door. I did not put any light on in the house. I assumed they were wolves, because I had heard them howling a couple of weeks before. Theirs is a different sound from coyotes—a long, drawn out howl.

MICHAELA WELTI

Our neighbour, young John van Loon, had a trapline, and he said he had seen wolf tracks in the snow. Never in all our time in Pemberton had we seen wolves come to that part of the valley. I woke up Max and told him to come and look. They were still there, with Rex alternately barking and whining. Max put on his rubber boots and a jacket, took a stick, and went out after them. He called to Rex, and the dog ran to him. But the wolves were not easily chased off. When Max chased them, they ran off a little way, and then sat down in a semicircle, facing him. Max drove them further off, and they finally disappeared into the forest. I think they would probably have devoured our Rex! Max remembered that when he was working in logging on the other side of the river some time before, he had seen a black wolf pass by when he was eating his lunch. When Rex got older he got painful arthritis of the hip, and we had to have him put down. Now we had no more guard dog!

Swaying Head—Not a Good Sign

This was just before we sold the farm in Pemberton. In the fall, the bears came close to the house because we had apple trees there. I could usually chase them away, but this one was slow to go, turning to face me, and swaying his head from side to side. There was also another younger one, but he did not come close, just stayed grazing in a red clover field every day.

One day, a friend, Jack, came by. He was a hunter from Whistler, and we told him about the bears. He said he had a licence for a bear and would come one day, toward the evening. He came, and we waited in the kitchen while it got dark. This was a bold bear and not at all shy. I told Jack about the swaying of the head and he replied, "This is the first sign of getting mad, getting aggressive." I was shocked, and realized it was actually dangerous to have such bears around. I opened the door a little, and saw he was under the

110

apple trees, munching apples. "He's here!" I said to Jack and Max. Jack got his rifle ready and went outside. After an ear-splitting shot, I saw the bear dash away into the bushes. I thought Jack must have missed him, but Jack said, "No, I got him right through the heart."

Why he had dashed away? I wondered, but Jack told me that shot through the heart, the bear will run perhaps ten to fifteen feet and then collapse, dead. After waiting a while, we went outside, and, indeed, the bear lay there, dead, sprawled out amid the ferns. The next day, Jack came with a truck to haul him away, with Max helping him load up with the tractor. Jack wanted the fur, the hide of the bear. Later on, he also shot the smaller bear. More people down the valley had trouble with bears. There was a young family with two small children who lived in a house trailer. One morning, they awoke to see bear-paw prints on the windows! They were shocked! Some other people owned a cabin further up the valley from our farm, and liked to spend a few days holidaying there. A bear came and ate their dog! They never came again, and were so scared they sold the property. This happened before we sold our farm in Pemberton Meadows.

Nuisance Deer

And this was not the end. The deer started to be a nuisance, too. One day, Max shot a deer that was eating my flowers in the rock garden close to the house. His intention was just to scare the deer away, since it was not shy and wouldn't leave. It ran away, over the road, but Max found it later, dead in the ditch. I was wondering if maybe some people had fed the deer, and if that was why they were not shy.

When we bought the Edgewood Farm in 1992, we had lots of deer around the house, and grazing everywhere in our meadows.

111

MICHAELA WELTI

Some stray horses came and also stray cattle that the owners had neglected to fence in properly. There was an old barbed-wire fence around the property, but this proved no problem for the animals. Actually, I think the deer got mad at us, that we claimed to own the property after they had had free access to it for so many years! They snorted at us and stamped their feet in protest! Max had to deal with a stray horse that got entangled in the old barbed-wire fence. It fled, snorting wildly, but never came back again.

There was another experience with deer, which I will relate. When we moved to Edgewood in 1992, Max wanted to continue with his system of building up topsoil with green manure. He had worked a few years on our field across the creek and the soil was starting to become very good. He seeded out a pasture mixture of clover, alfalfa, and grass seed early in the fall. The pasture came up nicely, but attracted more and more deer wanting to graze on it. When the snow came, we thought they would stop coming, but they didn't. They pawed away at the snow and kept munching right down to the roots. Sometimes, there were as many as twenty-five of them out there. Now they were even digging out the roots! Max was disgusted at seeing all his work going to waste. What was so special about that field? Why did they not go to the other pastures? There was lots of other land and fields they could go to. The following year Max put an eight-foot fence around the whole farm except for the woodlot.

Grizzly Encounters – Yes Plural!

Photo of Grizzly pastel
Photo credit to The Edge. Artwork by Michaela Welti.

One year, Max had seeded out a nice crop of carrots in the lower field by the creek. In the fall, the crops looked promising—but one day, we noticed some carrots had been dug up, but not eaten, as if some animal were trying them out. A bear? We had experience from Pemberton that sometimes a bear would come around and take a liking to carrots. Then we had to put up an electric fence. I said to Max, "Whatever it is, it will be back!"

So we watched every day. One day, we had just come home from a shopping trip to Vernon. I went down to the garden to get some corn cobs for supper, only to find that the cobs had been neatly eaten off, not even a kernel of corn left. I wondered, *What kind of animal would make such a clean job of it?* But then I saw a pile of bear scat and knew it was a bear. "He will be back!" I told Max. We had a dog by then, a black, short-haired Labrador named Benje. The next night, Benje began barking furiously at something we knew must be a bear. We went out with a big flashlight, and what we saw was a bear with two small cubs. A mother bear! She was light in colour with long hair, and the two cubs had even longer hair. They looked like sheep, and when they ran away, their long hair moved in waves over them. It was hard to believe that they were bears! They looked as if they had come from another world. Mystified, we went back into the house.

Later, Benje started barking furiously again.

I said to Max, "You take the flashlight and see what's going on." Then I heard loud roaring coming from the garden. Max came running into the house, slamming the door behind him.

"She stood up and was after me, but the fence was between us—otherwise, who knows what would have happened to me," said Max.

She must be a grizzly, I thought, and that proved to be true the next night. I had heard some banging noise from the chicken barn, and there was Benje barking furiously again. Max and I went out with the flashlight. This time, she had cleverly opened the feed box lid, leaning it open against the wall, and was sitting on the grain sacks in the box, munching happily away. Now we could see her clearly, her big round head. The hump on her back proved she was a grizzly.

The cubs were around, but had disappeared because they did not like the light shining on them. What were we going to do? Max was going to get his heavy gun. I was supposed to shine the light on her while he shot her. I realized how dangerous this was.

FROM THE GROUND UP

What if he didn't kill her? Just wounded her? She would attack us both! And we were outside the fence. I had another idea. "Go and get the pickup truck! At least you will be protected in the cab," I told him. "Shine the high beam on her and chase her away!" He did that, but she saw him coming and jumped out of the box and turned around the chicken barn into the field, with Max following. I heard his wheels spinning. He was stuck.

Then I heard the sound of him driving further down toward the creek, and I went back into the house. After a while, Max came back. He had kept on chasing her, but she would always rear up and try to confront him. At a certain point, the truck would not drive further, and he realized that she must have been under the engine compartment. After a while, she must have rolled out from under there and run away with her cubs. We would have liked to catch a glimpse of them in the daylight to see how they really looked. They seemed so beautiful and mysterious.

Years later, we had another encounter with a grizzly. It was fall again, a morning around the middle of October. I usually went out to feed the chickens. I had raised a flock of heritage chickens, an older breed. They were coming along nicely and, being five to six months old, they would soon be laying. On this day, I was shocked to see the entrance door split apart and pushed into the chicken house! There were no chickens around—only the young rooster lying dead by the wall. What had happened?

The bear had torn holes in the chicken wire to try to get at the chickens, but the small door where the chickens went in and out was too small for the big creature. The door had a round doorknob, which he could not open, so he had broken down the door. Max had disked an area behind the chicken barn, and there in the soft ground we saw the bear tracks. We could see they were definitely those of a grizzly. One could see the claw marks and the toe-prints clearly in the fresh earth. What was remarkable was the split frame

115

of the door. It took some power to split that wood! This time, we were not going to handle the situation ourselves. We phoned the wildlife conservation office in Castlegar and explained the situation. They said they would come and set a trap.

The trap was like a heavy barrel (a section of large culvert pipe) with heavy wire around the open end. One end had a trap-door that could be set with some bait. When the bear goes inside and pulls on the bait, the door slams shut and the bear becomes trapped. He came again the next night, but ignored the trap. The conservation officer threw some apples in there, along with some sardines out of a can. And he put some carrots in a can and hung it from the bait-hook. He was still unsure if this bear was a grizzly. The same bear had apparently bothered some people further up the hill from us. He was after their chickens, and had even tried to lift the roof off their chicken coop. He seemed to eat only chickens.

I kept in touch with this neighbour, and told him the bear did not go into the trap. He said he would butcher a rooster and bring it for bait, and he came and hung up the rooster on the bait-hook instead of the carrots. This did the trick! When the bear came again the following night, he went right into the trap. I could hear the bang as the door fell down. He started to scratch at it to get out right away, but he could not.

The next morning, we went and took a look at him. He looked like a black bear with a brown face. But my! He was fierce. He growled and hissed at anybody who came near. Lots of neighbours came to have a look at him, but some had little children who got frightened and ran away crying. He was so violent in his efforts to get free that I was afraid he would upset the cage. But the cage was well anchored to the ground and held firm. We had to wait all day for the conservation officers to come from Castlegar to pick up the bear. We were wondering how they would transport him to

FROM THE GROUND UP

Castlegar in the back of the truck, with him trying to wreck the cage, but they said they would tranquilize him first.

This was quite a procedure. The officer loaded up a long needle and fastened it to the end of a long metal pipe that Max got from the barn. I ran into the barn because I did not want to watch this, or to hear the noise he would make. He made a lot of noise all right, and bent the metal pipe with the hypodermic needle. He slowly sank down, but still kept on looking at us in a trance. The officer gave him another shot with the needle, and when he was really out, he opened the door and examined him closely. It turned out he was a grizzly all right, and a male. The officer said his orders were to shoot a grizzly male, but to relocate it if it was a female. So he took out his revolver and shot the bear through the back of the head. We said, "Thank you for helping us get rid of that beast!" And off they went. Later, he phoned us and told me the bear was about fourteen to fifteen years old, and weighed 400-500 pounds.

There were a few chickens left, but no matter what we did, they would never go back into their barn. We had to pluck them out of the trees when they roosted at night. They were still in shock!

Thinking of the stories I heard and have read about grizzlies, I must say that these bears are really powerful, majestic, and dangerous, but also beautiful in many ways. A creature that fits so well into the wilderness of Canada. If they would only stay there! Why have so many bears been encroaching on human habitats in recent years? Don't they have enough food out there? Are we or they increasing in numbers? Are they getting addicted to our food and garbage?

5. FELTING: AN ANCIENT CRAFT REDISOVERED

Felting the Natural Way

There came a day in mid-1985 when Max said, "No more kids coming home from school!" Indeed, we were on our own again after so many years of having kids around the house. It was a very emotional time for me, and I shed some tears. I knew that now a new chapter in my life was beginning. But I could not succumb to the empty nest syndrome. I had my artwork, which was varied and included spinning, knitting, weaving, ceramics, painting, and music.

My painting was specialized in soft pastel work, and I had already sold quite a few scenery pieces. I also taught evening classes at the Pemberton High School in ceramics and pastel work. I wanted to get out more and meet people. So I joined the BC Women's Institute in Pemberton. This was a good thing, because we did crafts, which we sold at bazaars, and once every four years, went to craft conventions at UBC where there were many delegates from all over the province. It was interesting for me to learn how an organization was run—things like reading the minutes and *Robert's Rules of Order*, and writing reports. I also enrolled in an art course by correspondence with a college in Minnesota. Through this, I won a prize in the beginner's category, with a pencil portrait of a young girl.

Mountain wildflowers.
Photo credit to The Edge. Artwork by Michaela Welti.

FROM THE GROUND UP

Portrait of Michaela's granddaughter.
Photo credit to The Edge. Artwork by Michaela Welti.

One day, I visited a friend in Pemberton, Mrs. Uri. She was of German origin, and taught fibre arts in Vancouver. She showed me her portfolio of samples she'd made from spun silk and wool. Among them were small squares of felted wool inlaid with flowers. An idea struck me at once: natural fibres like wool, silk, angora, mohair, and even linen had always fascinated me in their natural, unspun state. When they were spun into yarn, their texture and appearance would change. Yes, yarn was very useful. You could knit, weave, and crochet with it to make functional items. And I

MICHAELA WELTI

liked the beauty of the natural flowing fibre very much. Now I saw a way to capture that beauty with felting.

I went into a yarn store and asked if they had any books on felting. Felting wool was not so popular then. They had only a small booklet, handwritten, with very basic and easy-to-follow directions and sketches. But that was enough to get me going. I started to experiment with 12x12 squares, laying a thin layer of wool fibres out over a square of fine netting, one layer horizontal to the weave of the netting, then another layer diagonally. My aim was to make a thin felt with silk fibres in between. It had to be wearable. For the fibres to felt, the strands of wool or other fibres have to be laid over each other crosswise. Then I would take another fine netting square and lay this on top, and baste the netting squares together with big stitches, using sewing thread. This way, the wool fibres would not move during the felting process.

I dissolved some Ivory Snow soap flakes in boiling water and sprinkled the soap solution over the squares until it soaked in. Then I would start squeezing and pressing the squares with the flats of my hands, wearing rubber gloves. Sometimes, a rough surface like plastic bubble wrap is good to aid the process. Press with the flat of the hand, and move the squares around; turning them helps the fibres to felt. Then I lay the pieces flat in a sink, and pour boiling water from a tea kettle over them. I press the water from the felted squares, and see and feel if they are felting together. When they are, I give them a final rinse with very cold water.

Next, I cut the stitches and remove the felt from the netting. My first experiments turned out fine. Even the silk fibres had felted into the wool! Now I had more ideas: I made a few more felted panels combining hand-spun angora or silk yarns into them, with the idea of adding these to sweaters, matching the colours. The panels were just like watercolour paintings added to the sweaters. Landscapes and flowers! Amazing! In time, I improved my technique a lot.

Through the years, I had accumulated a lot of different fibres—silk, mohair fleece, finest merino wool, and cashmere. And I did a lot of my own dyeing with textile fast dyes. From these, I made scarves, vests, and sweaters, combining matching-coloured yarns of angora, silk, mohair, and wool.

Heidi Welti, 1987.
Photo credit: Welti Family photos.

MICHAELA WELTI

Rhonda Bampton, 2017.
Photo credit: Frank Appleton

Thirty years of felting 1987–2017.
Photo credit: Frank Appleton

FROM THE GROUND UP

The European magazine *Anny Blatt* was often my inspiration. Later, I acquired another book on felting by a Scandinavian author, which is now out of print. In it, there was a very interesting historical account. About 1980 or 1990, there had been a discovery in the permafrost of the Altai Mountains in Asia. It was a grave of a young woman of noble descent, dating back about a thousand years. Perhaps she had been a princess. Much of her clothing was in very good condition, amongst which they found very fine felt made with an unknown fine fibre. It was so fine—possibly from an animal that had become extinct—that experts could not identify it. Some thought the legend of the Golden Fleece might have originated from the material made from this fibre.

The Mongols are still using wool felt today to make their yurts, which are made from panels of felt. Since they were nomadic, they could disassemble the yurts to move them to new places with better pastures, and set them up again quickly. They had flocks of sheep, which provided them with milk, wool, leather, and meat. I believe that in times long past, there existed great cultures, especially on the European continent, whose great wisdom was based on the use of natural things. Today, we are only beginning to know exactly how the great pyramids of Egypt were built, a process that involved water, and the annual flooding of the Nile River.

My intention in using natural fibres in felting was to create sophisticated and beautiful wearables like vests, sweaters, scarves, capes, and jackets, along with cushions, wall hangings, and so forth. The fibres inside the clothes had to be very fine, like silk to line the wool so as to avoid itchiness. I made three sweaters like this, with my own handmade name tags. I thought about the price I wanted for them, put them in a black plastic bag and took them to Whistler.

125

MICHAELA WELTI

Of Sales and Thefts

By this time, Whistler had grown into a huge ski resort, with
many big hotels, shops, and restaurants. Tourists from all over the
world came to the resort every year. There was a boutique there
called Inge's. They had hand-knitted sweaters, knitting yarn, and
a small art gallery. I liked the store immediately. Inge herself was
a German lady. I went into her store with my black plastic bag
and took out my three sweaters. She was amazed! She had never
seen such felted panels combined with knitted items, especially
using angora wool, which was very expensive. She took them on
consignment immediately. I told her if I could get $150 each, I
would be happy. But the sweaters quickly sold for over $200! Inge's
commission was forty-five percent. So I kept on making more
sweaters to sell at Inge's. She told me later, "I will never forget
how you came into the store with that black plastic bag and took
out those wonderful creations!" And she told me where my sweaters had gone after they were sold: New York, Germany, England,
Switzerland, and many other parts of the world.

Around Christmas, when the store was full of people, two of
my sweaters disappeared without being sold. Inge said there were
so many people that she couldn't keep her eye on things. More
theft occurred after that, and it was very sad. Luckily, Inge had
insurance, and I did get my money for the stolen items. One of
my sweaters went to Montreal and sold for $600, but I had a hard
time getting my share of the money, $300. Later, I lost two more
sweaters: one was a first-place prize-winning sweater in a magazine contest.

I learned to be more careful with selling my creations or leaving
them for stores to sell on commission. Inge's Boutique eventually closed. After we moved to Edgewood, I made an attempt to
sell some of my felting creations and hand-woven tablecloths at

FROM THE GROUND UP

the Vernon Art Gallery Gift Store. Again, however, I experienced losses and disappointment.

Unique in Every Way

My creations were highly original and no item was like another. Later, I found that I could do better in selling my work by going to craft fairs and selling items directly myself. There is much fascination with felting. It is so unique. There is no spinning, weaving or knitting involved—just fibres of wool, silk, alpaca, cashmere, angora, dog hair, qiviut (muskox), or mohair (goat). One can also knit wool yarn with large-sized needles and incorporate the fibres on both sides, and so make a design. The felt can also be rolled with a thick dowel, back and forth between bubble wrap. Using the roller dowel on a bamboo mat makes for a flatter felt. One can also incorporate linen fibre, lacework, or nylon netting into the process, as the different fibres will felt into the background wool felt. And there are now felting needles available in yarn stores. With these, you can sculpt animals or other designs on a surface of wool fibres or knitting, which eliminates the need to stitch netting to hold the fibres in place.

With experimenting and imagination, there are many possibilities to the art of felt. One is placing a design on both sides of a vest or shawl. There is also the advantage that, if one cuts felt with scissors, unlike fabric, it will not fray. Also, holes can be made in felt without fraying. And of course you can sculpt designs with fibres, giving it a three-dimensional effect. With this technique, living in Edgewood, I started to make wall hangings with images of birds, flowers, leaves, and hummingbirds, standing out from the background. The "sculpted" flowers look like real flowers. (Photos of some of these creations are available.)

127

When it comes to ceramics or pottery, I prefer hand-building to the wheel. Hand-building and sculpting in clay is so much more versatile and original. When we moved to Edgewood, I bought a kiln. I also had a bigger loom that produced sixty-inch wide fabric and had flying shuttles, which speeds up the weaving. As I did in Pemberton, I taught painting, ceramics, and felting in the fall and winter in Edgewood, with a small group of five to six people. With the sized felting needles now available in the wool-craft stores (called dry felting), it's possible to extend this art.

Pottery
Photo credit: The Edge

6. EDGEWOOD

A Unique Place

Photo of Inonoaklin Valley by Jeff Holman.

Before we sold our farm in Pemberton, a friend of ours told us about a nice green valley somewhere in the Kootenays, close to a lake. He remembered this as a place he would choose to retire. We didn't think of it any more until we had settled down in Edgewood.

MICHAELA WELTI

Every time we came home from the Okanagan Valley, driving over the Monashee Pass and coming down into the Arrow Lake valley, we enjoyed the special feeling of arriving in a unique place. You descend into the valley, with a backdrop of high, snow-capped mountains. This is the Kootenays. About ten kilometres before the Needles Ferry crosses Lower Arrow Lake, there is a junction on the right-hand side of Highway 6. It leads into the Inonoaklin Valley, ending ten kilometres later at the village of Edgewood on Lower Arrow Lake. It is a lovely green valley, with fields of promising fertile soil. Farms were spread throughout the Inonoaklin Valley.

The Edgewood Women's Institute had published a book: *Just Where Is Edgewood?* It was a good title, since many people who had grown up in BC had never heard of this secluded valley. In this book, we read about the history of Edgewood and the Inonoaklin Valley, which we found very interesting. The Inonoaklin River starts from the Monashee Mountains and winds its way down through the valley, plunging over the falls into Arrow Lake. Near the falls is where the village of Edgewood is located. "Inonoaklin" is an Indigenous word meaning "winding river." In the early 1900s, many settlers came from many different countries: England, Holland, Germany, Scotland, Ireland, Russia, Ukraine, and, more recently, the United States, as a surge of Americans evaded the Vietnam War.

Mr. Gerber, a cousin of the famous Swiss cheese Gerber family, started a cheese factory in the valley in 1930. A cousin of the famous Dr. Banting (who discovered insulin with Dr. Best) kept the general store in Edgewood. The village was situated at the shore of Arrow Lake, and was then bigger than it is now. The valley had a heavy growth of timber—cedar, fir, pine, hemlock, cottonwood, and birch. Some of the cedars were six or seven feet in diameter. What a job it was to clear those trees to turn the land into pasture! The settlers logged and cleared the land, but in 1925,

130

FROM THE GROUND UP

a huge forest fire raged through the valley, and then the valley got the name Fire Valley.

In 1964, the Columbia River Treaty with the United States saw the Arrow Lake valley flooded, as dams were built for hydro-electricity and flood control. The whole Edgewood town site had to be moved to a bench on higher ground, and new houses were built. They built a new legion hall, an outpost hospital, a post office, a credit union, a new school, and a general store. The Red Cross Outpost Hospital was sponsored by the BC Women's Institute and is still operating in 2017 as the Edgewood Community Health Centre. It is administered by BC Interior Health. This is most appreciated, as the nearest hospital is one hour away in Nakusp, or two hours away in Vernon, via the Monashee Pass. The health centre has a nurse in attendance five days a week, with two doctors from Nakusp alternating twice a month.

Before the road was built over the Monashee Pass to Vernon, sternwheeler steamboats were used for transportation up and down the Arrow Lakes. Besides logging of beautiful timber, the Inonoaklin Valley got a reputation for quality farm products. Apples, cherries, strawberries, grain, hay, alfalfa, plus sweet clover were grown with great success. Raising cattle was also popular. When we came in 1992, there was just cattle raising and one dairy farm. There were no commercial crops grown that required the application of pesticides. This suited us fine, as we wanted to grow organic crops.

MICHAELA WELTI

Settling into Our New Home and Farm

Inside of our log house.

Max started work on the house, as there was much to be renovated. I got busy with the garden, building a rock garden at the front of the house. I joined the BC Women's Institute in Edgewood, and made new friends. We soon introduced ourselves to all our neighbours. People were very friendly and easy going—not at all competitive like in the Pemberton Valley. In time, many of our friends from Pemberton came for a visit and to see the farm we had bought. They admired the log house and the area.

Log House in Edgewood.
Photo credit: Welti Family photos.

FROM THE GROUND UP

Not far away, up in the hills to the north, was Whatshan Lake, which was great for swimming in the summer, and had nice beaches. Due to the seasonal rise and fall and the use of the water for hydro power, Arrow Lake did not have good beaches, and the water was colder there. But when the lake level was low in the late winter/early spring and the beaches were revealed, you might find Indigenous artifacts like arrowheads ("Arrow Lake") from the time when Aboriginal peoples had travelled along the Columbia River. We found attractive stones there, too, which we used in the rock garden.

In 1993, Max and I went to Switzerland for a reunion of the Welti family. We all gathered together in Interlaken, which everyone enjoyed. Max's brother Karl had immigrated to Canada in 1980 and bought a hunting lodge on Cariboo Lake. It was a dream fulfilled for Karl, as he was very fond of BC. Today, they have retired to Likely, BC, on Little Lake.

Green Manure—the Way to Grow

Max wanted to do grow a few crops like carrots, grain, and hay, and, most of all, to try out his organic soil-development method. We also bought forty acres of land in the valley, separate from our farm. This was pasture land with no house or barn. We joined the Okanagan Certified Organic Association, which was just beginning, and went to Vernon for the monthly meetings. There, we met other certified organic farmers—of which there just a few at that time. We visited some of the more established of them. There was the Wild Flight farm in Mara and the Pilgrim farm in Armstrong. It was interesting to see what they were growing.

Every year, there were farm visits planned. These were day trips with lunch, and the owners would show us around their farms. Some of the inspectors would instruct us in how to grow certain

133

crops and what the specifications were that we should follow. We made good friends with them. Every year, we paid a fee of $400-$500 for inspection of our crops.

In time, we found that living in Edgewood was something of a drawback, because we were so far away from the markets in the Okanagan. However, Max grew some carrots and potatoes. He developed the soil with green manure, using mainly buckwheat, peas, or orchard grass. The forty acres we had bought on the valley bottom had more clay and silt. One field close to the river needed to be worked up more with green manure than the rest, but it was a beautiful sight to see a whole field of buckwheat, with their white flowers standing four to five feet high. One year, Max grew ten to twelve acres of hard wheat, which became a very good crop. We went over to Armstrong and bought an older grain combine that worked just fine, and then took a sample of the wheat to Rodger's Grain Mill in Armstrong to be tested. When we came home, there was a phone call from the mill. They said our wheat had twelve percent protein, and that they would buy the whole crop.

Our hay crops also improved. One year, Max had seeded a field of red clover by the main road. It was so nice and red when it bloomed that people commented on it, saying that it had been a long time since anybody had grown such crops of grain or clover in the valley. For a few years, Max sold hay crops and carrots. If we had a surplus of vegetables, we would sell them locally, or at the farmers' market. Eventually, we sold the separate forty acres to a local farmer. Max had proved again that his method of organic green manure soil development worked.

He believes that it would work on all kinds of soil around the world. We cannot keep taking from the soil without building up the topsoil. We must look to the forest with its beautiful majestic trees. How do they sustain themselves? They are regenerated from their decomposing leaves, and also from the bushes, grasses, and

FROM THE GROUND UP

ferns growing around their roots. The wild fruits also feed the wild animals and birds. It doesn't matter how big your farm or garden is—always set aside a third or a quarter of it for green manuring, to build up the soil.

There are different ways of building up the soil. One can work with only mulch, like grass or leaves. Grass and hay are only good if cut before they go to seed, because the seeds will sprout under the mulch. Legumes and clover used as green manure will put nitrogen into the soil. Max discovered that if he let the buckwheat bloom for a while, seeds would be produced and after he disked in the crop, the buckwheat seeds would sprout again for a second crop. But this crop is not frost hardy. There are now very good organic fertilizers available, like alfalfa meal, kelp granules, guano, worm castings, rock phosphate, and glacier rock-dust, which we would recommend, especially for fruit crops. These fertilizers have many minerals, and a good application every three years works well. And of course we make a compost heap! If you have a good stand of comfrey somewhere, you can cut it down three times a year and add it to your compost heap or use it as mulch.

Ruth Stout invented the "permanent mulch garden," and became famous with her book, *The Mulch Garden*. Even back in Switzerland, I read about her. There, the book was called, *Der Matratzen Garten*. She lived to be ninety-six years old, still eating her vegetables from her garden. She did not plow or dig her garden anymore. She got old hay bales every year from farmers and replenished her mulch garden with these. Her method was just what nature has been doing for thousands of years.

New Friends

When we first moved to Edgewood, it took us several days going back and forth to Pemberton until we were finally settled into our

MICHAELA WELTI

new place. When we returned one time, there was a container with beautiful salad plants on the doorstep, ready to be put into the garden. But there was no name on the container, so we had to find out who had left it! We asked our nearest neighbours across the road, but they didn't know. As it happened, we were ready to visit other neighbours and introduce ourselves anyway. We started with some who lived up on the hill, across from our fields, in a very original self-built house—Jean Bassett and Frank Appleton. Jean had been the one who'd brought the plants. We became good friends, with a mutual interest in gardening and plants. Jean showed me around the area, and we would find wild nut trees and blackberries, and dig out old roses from abandoned gardens on the Arrow Lake shore. We also found some wild asparagus and clay for pottery.

Jean was a teacher in Nakusp at that time. One summer, we grew a lot of flowers for drying. We arranged them into wreaths, swags, and decorative items, and that fall before Christmas, we took them to the Christmas fair in Vernon. I also took some of my felted creations. We were successful, but there was a big snow-storm over the Monashee Pass, and we had to creep home slowly. We arrived safely.

In the following years in Edgewood, we made many more friends. Across the lake in Fauquier, we met some German and Swiss people. We held yearly steak dinners for seniors, square dancing, and family reunions, and celebrated in the apple juice building, which Max had renovated into a guest house. I started to give art classes in the winter months in painting, felting, and pottery. As we both grew older, we slowed down a bit. I tried to simplify my garden, because in summer when it was so hot, Max had difficulty irrigating so large an area. But we still maintained our vegetable garden, as it was important to be self-sufficient

FROM THE GROUND UP

through the winter. We saved a lot of money that way, and it kept us healthy.

In time, Max realized he had too much equipment, so he planned an auction, which he advertised on the internet. When the day came, we were surprised to see so many people. The auctioneers came from Vernon and people came from Lumby and even Pemberton to buy. Max sold everything he wanted to sell, and with the proceeds, bought a new, smaller Kubota tractor with some attachments for the garden. After we celebrated Max's eightieth birthday with a family reunion, Max wanted to visit his relatives in Switzerland. I stayed at the farm, since it was September and there was much to harvest from the garden. We had grown peas, beans, and flax for drying, along with some hulled barley, all of which needed attention.

Addressing Max's Health Issues

When Max was in his early eighties, he started to have some health issues. His bowels were not functioning well. His doctor recommended a colonoscopy, and they discovered a tumour in the rectum. The doctor said it was cancer, and they kept him that day in the hospital. I went home shocked. I let the kids know, and our sons, Mike and Alex, came right away to see their dad in the hospital in Vernon. They talked to the doctors.

We had lived a healthy life on the farm, and I was wondering how this could have happened. I asked the doctor how long it takes for such a cancer to develop, and he said twenty to thirty years. Looking into the past, I discovered that Max had worked at a vegetable farm in the Fraser Valley at that time where they had used a lot of insecticidal sprays on vegetables, and chemical fertilizers in their fields. I blamed the cancer on that, and on the spraying of the potatoes and turnips in Pemberton.

The doctors explained to me and my sons exactly what the treatment would be. They were not going to operate right away, but would try to isolate the tumour, which was already a size three or four. Max had an operation to open a side outlet for the colon with a bag outside, then his treatments started. Luckily, we had a friend staying with us for a while, so I could go with Max for his treatments in Kelowna. They have a cancer centre in the hospital, and a lodge where cancer patients and family can stay during their treatments. We stayed two days in the lodge and moved to a motel near the hospital, which had a kitchen where we could cook our own meals. The treatments would last five to six weeks, with chemotherapy and radiation. Max was able to walk every morning to the hospital for his treatments.

It was spring—April and May—a nice time, and we were optimistic about the outcome. We could go home on weekends and tend the garden. The cancer centre was decorated with lovely artwork. The entrance hall was big and also decorated with beautiful art. One day as I waited for Max to have his treatment, I was observing all the people who coming for their treatments, some in wheelchairs, some in wheeled beds. With some, one could see how the cancer had progressed already; their faces were so white. I felt a deep compassion for these people, and, somehow, I knew there was something wrong with the way we were living our lives. That night, I experienced a spiritual event. I prayed for all those people and for Max. I knew from the bible how Jesus had healed so many people because they believed in him! There was the story of the woman who only touched the hem of his garment and was healed. I realized there is more needed than human science, medicine, and invention. We need a living faith in God who is there, and created us and the universe.

The next morning, I felt peaceful and certain of something I could not describe. When Max came home that day, he told me

FROM THE GROUND UP

the doctor had examined him to see if the tumour had responded to the treatments, and had found that it had shrunk to half the size. We were optimistic, and finished the remaining weeks of the treatments. There were some side effects from the radiation visible on his feet and his hands, but this went away after a few weeks. Next, the big operation was done, to remove a piece of the colon where the tumour had been. If everything went well, they would stitch the colon up to his anus again, so he could have normal bowel movements. Everything went fine! The doctor said almost nothing was left of the tumour. We were so thankful to God who was there to give Max a new life. In August, following the operation, we celebrated the event with a family reunion at our son Mike's house on Shuswap Lake.

Now, seven years later, Max's cancer has not reoccurred, and he has just celebrated his ninety-first birthday. His health is excellent, and he is able to tend the garden and is even falling trees for next year's firewood. His health remains normal and he is not on any medication.

The Fire

One more notable event took place, this on December 15, 2012. It was in the morning as I was preparing dough for bread. Max was in the basement. We were just about to have a cup of coffee together when I smelled smoke. I was in the kitchen, and I saw smoke curling along the ceiling. A few weeks before, we had installed a new stainless-steel chimney, because the old chimney had not been functioning so well. Anyway, the fire started in between the ceiling and the roof. We had no fire insurance. It spread very quickly and, in two to three hours, it had burned the house down, right into the basement, even despite the volunteer firemen arriving with their water-tanker truck. I had to drive our pickup out of the garage

139

attached to the house, and run into the house to rescue my wallet, which was on top of the kitchen cabinet. To do this, I had to take a deep breath and hold it, because the house was filled with smoke. Max, meanwhile, was trying desperately to start a water pump out in the barn, to pump water from our creek, but he couldn't start it.

The only things Max and I were left with were the clothes we had on. We were lucky to have our guest house, the former apple juice factory. We had three queen-sized handmade log beds that Max had made, sheets, towels, dishes, tables, and a big wood stove. That day, many people came to help with food and clothing. They even brought us firewood for the oven, because our wood supply by the house had also gone up in smoke. Our beloved log house had gone so fast, we were really in shock for a while. Max had done so much work renovating it! It was our pride and joy! Completely gone in such a short time! But maybe it was good that it burned so completely. What would we have done with a half-burned log house? It would have been almost impossible to repair. We had the site cleared and covered with soil. Today, one cannot see that a house once stood there.

We did not grieve too much over our losses. They were just material things. We decided to stay on our property, and hired our son Alex and his son Adam to renovate the old apple juice factory that became a guest house and was now already a new home for us. We added a bathroom, new floors, a new kitchen, a living room, and, lately, a fully enclosed veranda, which we enjoy very much. It's a convenient place for us at our age. We still have a guest suite with two bedrooms and a kitchen, with a toilet and shower for visitors and family. The big shed that Max built when he was seventy years old is still standing and serves us for a workshop and garage. It's a home for the tractor and its implements. Housekeeping is easier for me now. Everything is on one floor.

7. OF HERBS AND HEALTH

Teas for Healing and Health

It was natural for me to be interested in herbs. At an early age in Switzerland, I had experienced the fields and hills of the Alps that were full of wild herbs like enzian, arnica, yarrow, Lady's Mantle, eyebright, thyme, dog rose (rose hip), camomile, plantain, hawthorn, buckthorn, nettles, wild raspberries and strawberries, elderberries, and many more. We had one herbal book at home from which I learned the meaning and use of herbs. My mom was also interested in herbs and health. We had excellent herbalists in Switzerland, like the famous Dr. Vogel, Reverend Kuntzle, and Dr. Bircher.

When I came to Canada, I continued to be interested in plants with herbal and healing properties. When we lived in Pemberton, we had access to the open-shelf library in Victoria. This gave me the opportunity to read many books on herbs and natural healing. What I found astonishing was that, in the history of herbalism, people in one part of the world would find the same use for herbs in healing as other distant people, even though they had no communication. Of course, those in different climates would discover

MICHAELA WELTI

other, different herbs. We had started to use herbs as part of our natural lifestyle, both in Pemberton and Edgewood.

Every year from about the beginning of May, we would collect fresh herbs for tea—the blossoms of hawthorn, young birch leaves, wild strawberry leaves and blossoms, raspberry leaves, dog rose petals, and buds. Later, there would be linden flowers, elderberry flowers, and berries. Dandelion and nettle shoots we would eat like spinach. There were wild violets in Edgewood and many hawthorn bushes. We would also find cascara trees, wild cherries, chokecherries, morel mushrooms, and hazelnuts. We would dry many of these wild herbs, and then in the fall, combine them with garden herbs like camomile, calendula, mint, lavender, bee balm, lemon balm, meadowsweet, and rugosa petals. We would have many different jars of tea that would last the whole year. We drank two cups each day. I never bought any tea from the store from then on.

The most influential herbalist for me was Maurice Mességué, a Frenchman born in Provence. His books became very famous in Europe. In his first book, *Of Man and Plants,* he wrote about how privileged he was to connect with many famous people in Europe, and how he was able to help them with herbal remedies. But this put him into conflict with conventional medicine, and he faced many trials and court cases against him. He won them all. In other books, he tells us how careful we should be when growing or collecting herbs in the wild. Never pick any herbs growing beside busy roads or highways, nor any close to sprayed fields or orchards. Never use artificial fertilizers. The plants will grow. But something changes in them and influences their healing power. Today, so many people are on prescription drugs. They do not use any natural herbs because of the intervention of conventional medicine. But why are so many people still sick today? Mességué had a different method with health and herbs: he prescribed hand and foot baths

FROM THE GROUND UP

in hot water that had herbs steeped in them, but not boiled. A kind of herbal osmosis.

Our daily bread in time became very important to us. We started to bake our own bread early on in Pemberton. We had bought an electric grain mill from Europe, and milled our own flour from hard wheat and rye. A few years ago, we visited the old grist mill in Keremeos, BC, which has been renovated as a tourist attraction. We watched a demonstration of flour milling the old-fashioned way.

But when he showed us how they sifted out the germ and the bran, I was devastated. They did this so that the flour would have more shelf life! Even the so-called whole-wheat and rye flours on the market today have the germ and bran removed. They sell these separately, as a health additive. This was one more reason to stick to our own method of milling and baking bread.

We persisted in growing a garden every year, with a root cellar to keep our potatoes, carrots, beets, parsnips, turnips, apples, and pears. Max made a box with a wire screen around to keep the mice out. He filled the box with the root crops and buried it in a hole in the garden, deep enough so that the frost could not penetrate. Then he heaped earth up on top. He left the box there until the snow disappeared in spring. When the box came out and we took off the lid, the vegetables looked and tasted as fresh as if they had just been dug out of the garden. They were better than from the root cellar. I do freeze beans, peas, broccoli, kale, and tomatoes. We have red currants, raspberries, black cap raspberries, blackberries, and strawberries, which we use for jams and pies. Sometimes, I buy a few organic vegetables toward spring if our supply runs out. But we save a lot on groceries.

I taught myself to make tinctures of echinacea, rosemary, and many other herbs. This has kept us healthy into our old age. (In

2017, I am eighty-four and Max is ninety-one!) Today, we are not on any medication.

Refreeze the Glaciers?

I would like to mention our observations on climate change here in Edgewood. In our early years, we had excellent gardens. In the years to follow, however, we noticed some disturbing things. We have had more and more dry spells in the summer. The creek on which we depend for our water, and for the garden, has been dangerously low in late summer. Temperatures have been hotter. With the passing of years, many birds, bees, and other insects have disappeared. There are fewer robins now. We could remember when there was a loud chorus of them at dawn. One year, nobody had any apples on their trees. Many fruit trees were infested with coddling moth, with the leaves curled and brown.

But one day when I went for a walk to the creek, I saw a wild apple tree in the bushes, with its branches hanging full of apples. I got closer and saw that the apples looked beautiful and disease-free, and that the leaves were shiny with not a speck of damage. It was like a miracle. Further along were more apples of a different kind—light green/yellow. They were not quite ripe, so we left them on the trees for two weeks more. Why had the coddling moth not attacked these wild trees? Nobody had planted them; they had seeded themselves. Nobody had pruned them or fertilized them. But they had enough water right at their feet. We took some pails and the tractor down to the creek, and Max pushed a way into the trees. From them we got enough apples to last us quite a while. What had happened? What could we learn from this? Is nature self-sustaining? Is this not a lesson?

I remember in years past, when we would go into the wilderness to pick huckleberries, black cap raspberries, high-bush cranberries,

FROM THE GROUND UP

raspberries, and mushrooms, both in Pemberton and Edgewood, they would be plentiful, with lots for man and wild animals. In the last twenty years, though, it has all slowly changed. Every fall, we get many bears roaming through our gardens for food and invading the garbage. They seem to be desperate, because they will eat even unripe fruit and unusual plants. Often, they are dangerous to humans and incidents have occurred. I have not seen any chipmunks for a while, and the chirruping of the squirrels seems to have disappeared. What is happening out there in our environment? Are Rachel Carson's predictions from *Silent Spring* coming true? In the past, there were many people roaming the forest looking for pine mushrooms, a delicacy for the Japanese, and people made money from them. Now, we don't see or hear of any mushroom pickers anymore.

We hear a lot about climate change but do nothing about it. The logging trucks and chip trucks are rolling without cease on the highways. We hear that the glaciers and Arctic ice are melting much faster than was thought. By 2030, the ice will be gone! Solution? They will try to refreeze the glaciers with powerful machines (CBC News, February 16, 2017)! This sounds like madness, but by that time, I think Max and I will not be around to worry about it.

We are thankful to be living out our lives in the little community of Edgewood. We can see the stars, planets, and the moon on a clear night, contrary to the big cities with their pollution and bright lights. We are self-sufficient and thankful to live in this country.

Storing Grains

I would like to mention something about storing grains, dried beans, lentils, and peas. We buy all these in large quantities, so it is important to store them right. The best is a cool, dark room for grains and flour. When we mill grain with our electric grinder, we store the fresh-milled flour such as whole wheat, spelt and rye in Ziploc bags in the freezer. Flour stored in kitchen cabinets at room temperature is liable to become stale. It is advisable to add a pinch of Vitamin C powder to the water when cooking rice, barley, or other grains. Vitamin C powder will detox any mould or staleness. Even dried pasta can get a stale taste if it is stored too long. (I make my own pasta fresh now.) All grains contain phytic acid, which some people become sensitive to. A way to avoid this is to soak the grains in water for twenty-four hours. A change occurs during early germination, in the course of which the phytic acid is metabolized and eliminated.

Additives and Preservatives

Some years ago, we stopped using the usual white, iodized Windsor salt, which has a lot more refined pure sodium. Now we use only grey pure sea salt or rock salt.

There are so many additives and preservatives in food out there, and many more are being added all the time. We have no idea what some of them are, or their effects. Just go into one of the big supermarkets and look at the vegetables and fruit: super-large salad heads, huge cauliflower and broccoli heads, shiny, large apples and tomatoes, massive avocados; even the mushrooms are getting bigger. Huge grapes, and raspberries, strawberries, blackberries, and blueberries are available now in the wintertime. One little plastic container that contains only a cup-and-a-half of berries sells for five dollars. It all looks very enticing, seductive, and magical. But what is behind all of this? I am very suspicious about it.

When our son Mike came for a visit a few years ago, he brought us a flat of big strawberries, which were not in season. We ate some

FROM THE GROUND UP

during their visit. After they left, we didn't eat them because we had the feeling that something was not quite right. We let them stand around for a while, but they did not rot, as is usual with soft fruits. After a while, I threw them onto the compost pile. After a few weeks, they had still not decomposed! Magic!

What is happening to our food? Things have become a lot worse. Now we have genetically modified organisms where they are able to transplant genes and pesticides into plants and animals. Despite the growth of the organic food movement, commercialism and the almighty dollar seems to win in the end. I fear for our heritage plants and seeds. We will have to do what we can to save seeds and live a more natural life.

I will close this chapter with my recipe for our daily bread recipe.

Our Daily Bread Recipe

First, make a sour-dough starter:

Take 2 tablespoons of fresh-milled rye flour. Mix in a small glass container with enough water to make a soft dough. Let it stand at room temperature for 24 hours. Cover lightly with a cheesecloth. The next day, add 1 teaspoon rye flour plus a little water to keep the dough soft. Let stand for one more day. On the third day, add another teaspoon of rye flour and let stand for one more day. After three days, the dough will have developed yeast and bacterial growth, and should smell sour. It should have bubbles and have increased in size (risen). If you are successful with this sour-dough mixture, put a lid on your container and keep it in the fridge.

I work my dough for the bread in a KitchenAid mixer:

Take 5½ cups whole wheat flour or other grain like spelt flour or rye flour, or one can divide the 5½ cups and make a blend of grains. Add 2½ cups all-purpose white flour, 2 scant teaspoons sea

147

MICHAELA WELTI

salt, and 1 scant teaspoon dried yeast. All together you should have 8 cups flour. Sometimes, I make a European spice mix: 2 tablespoons each of fennel seed, coriander seed, and caraway seed, plus 1 tablespoon anise seed.I grind this mixture in my coffee grinder and seal it in a covered jar. Keep in the fridge. Add 2 tablespoons of this mix to the 8 cups of flour.

The evening before you want to bake: Take the mixer bowl and add 2 cups of the flour mixture (from the 8 cups), and mix the sour-dough from the fridge into this, adding water until you have a soft dough. Then put a plate on top and leave it at room temperature overnight. This makes a sponge. The next morning, the sour-dough will have made the sponge rise and it should be full of bubbles.

Now start **a new sour-dough for ongoing use**: take 2 heaping tablespoons of this sponge and add 2 tablespoons of rye flour and 1 tablespoon of white flour, adding water to make a soft dough. Take a clean glass container (every time), let stand on the counter for the day to rise, then put away in the fridge as before for future use. **This starter keeps for up to two weeks in the fridge if you stir it after a week**. If you don't bake for a while, it can be frozen indefinitely. In this way, you won't have to start the sour-dough from scratch every time.

Back to the bread mix: Put the mixing bowl on the machine. On slow speed with the dough-hook, start to mix in the remaining 6 cups of flour, adding water slowly, by the spoonful. The dough should be neither too soft nor too stiff. Shift to a faster speed when you have added all the flour. After a while, the dough should become pliable and leave the sides of the bowl. Beat at a faster speed for 2–3 minutes. Take the dough-hook off and leave the

FROM THE GROUND UP

dough to rise with a plate or plastic on top. The dough should rise for 2-3 hours, depending on temperature. Scrape the dough out onto a floured counter and let it rest for 10 minutes. Divide it into two equal parts. Give each portion a quick kneading and form into loaves, round or oblong, as you wish. Put them on baking parchment on a wooden board and cover with a plastic sheet to rise for 20-25 minutes. Heat the oven to 425°F well ahead, and slide a baking stone or porcelain floor tile into the oven. When ready to bake, take your loaves with the parchment paper and slide them onto your baking stone in the oven. After 10-15 minutes, adjust the temperature to 400°F, and bake for 45 minutes. Instead of a baking stone or tiles, you can use upturned cookie sheets, which also work well.

Sounds complicated? It isn't, really. The machine does the kneading for you, and the bread is not sour tasting. The more flour you mix into your sponge the night before, the more the bacteria works through the dough, and the more sour taste is given to the bread. This bread actually tastes better after a few days. Sometimes, I boil up some millet or quinoa the night before, and add a cup to the dough mix. Or you can add a half-cup of flax seed.

8. SPIRITUAL VALUES

Looking back over the years, we are thankful to have lived in beautiful British Columbia, Canada, where we have freedom of speech and religion. Earlier in my story, I had mentioned that the bible had become the centre of our lives. Max knew much of the bible even before I came to Canada. The Word of God, he received with sincerity. Looking back over our lives together, we had many encounters with religious sects or groups that were interesting to us and about which we were curious. They all claimed to be Christians, though they all differed from the bible's teachings. Either they added to, or subtracted from, the bible, and formed their own Gospel, writing their own bibles.

It became very confusing to find a way with them. It was astonishing to see how many followers they attracted—thousands, millions of them. We found it necessary to read the bible with sincerity. To confront these sects and give them answers, we studied the bible more. After all, the bible that we have today (like the King James English version, and the new standard version) has prevailed for over 2,000 years. We found many answers for the different sects and also for ourselves; a living faith that we could apply to our own lives. To answer the different sects, we quoted

Jesus in Matthew 24:5: "And many will come in my name and say, I am the Messiah! and they will lead many astray." And Jesus also said, in Matthew 18:20: "Where two or three are gathered together in my name, I will be amongst them." So we did not need to join any sect or church. Today, the world has become a medley of religions, bringing great confusion. But even this state was foretold by the bible. Jesus said in Matthew 24:35, "Heaven and Earth will pass away, but my words will never pass away." The God who is still there, still has the whole world in His hands.

I would like to conclude my story with what I will call: "Art and the Great Artist."

Around the world, we can visit galleries to see and admire artworks of painting, sculpture, ceramics, and so on, from famous artists throughout history. Today, some of this artwork gets sold at auction for thousands and millions of dollars. Only the very rich can afford to buy it. At the beginning of the bible in Genesis, it says: "God created Heaven and Earth first." Then He said: "Let us make humankind, of dust from the ground, in our own image, according to our likeness." If we are made in God's image, we also have the ability to be creative. This creativity is a talent, a gift from God. He is the Great Artist. Art is a gift from God, and His artwork is the Creation, which He gives to us freely. Every time a human being is born, we see God's artwork. In Romans 1:18, it says that in the creation of the natural world, we can recognize the Creator.

Science claims that heaven and earth started with the Big Bang. Who started the Big Bang? In Romans 1:22, it says, "Claiming to be wise, they become fools."

What are we doing with God's creation, our own planet? Are we giving honour and thanks to His wonderful artwork?

Today, we have climate change, a result of pollution and poisoning of the earth from mining our resources.

Are we still able to do something about it?

ACKNOWLEDGEMENTS:

My utmost thanks to Frank Appleton, author of *Brewing Revolution* (Harbour 2016), for correcting and transcribing my handwritten manuscript and preparing it for publication.

And my thanks to Jean Bassett and Jeff Baker of the Edgewood Community Internet Society, for helping me with organizing and formatting to bring this book together.

CPSIA information can be obtained
at www.ICGtesting.com
Printed in the USA
LVHW02s1042230418
574462LV00003B/3/P